At the Crossroads of Science & Mysticism

PAVEL FLORENSKY

AT THE
CROSSROADS
OF
SCIENCE & MYSTICISM

On the Cultural-Historical Place
and Premises of the Christian
World-Understanding

Translated and Edited
by
BORIS JAKIM

First published in the USA
by Semantron Press
an imprint of Angelico Press 2014
© Boris Jakim 2014

For information, address:
Angelico Press, Ltd.
4709 Briar Knoll Dr. Kettering, OH 45429
www.angelicopress.com

Paperback: 978-1-62138-085-6

Cover image: *Delicate Tension*,
by Wassily Kandinsky, 1923
Watercolor and ink on paper,
Museo Thyssen-Bornemisza, Madrid
Cover design: Michael Schrauzer

CONTENTS

Translator's Introduction

PAVEL FLORENSKY wrote two major complementary works of religious thought: (1) *The Pillar and Ground of the Truth* (1914),[1] which he called a "theodicy"; and (2) *At the Watersheds of Thought: The Elements of a Concrete Metaphysics* (composed between 1917 and the first years of the 1920s),[2] a collection of writings which can be called his "anthropodicy," an attempt to vindicate the works of man before God. The lectures translated in the present volume represent the most substantial theological contribution in *Watersheds*.

Florensky variously defines his task in writing these lectures: "My task is threefold: first, to prove the thesis that our epoch is characterized by a discontinuity in scientific and philosophical thought and in culture in general (the characteristic feature of this epoch being a turning back); second, to prepare the ground for the premises of the Christian world-understanding; third, to emphasize my own attitude toward the freedom of thought and the freedom of culture."[3] "We wish to show the existence of a shift in the spiritual existence of the world, a shift in the direction of a Christian world-understanding that will establish concepts extremely favorable for the Church. People who had departed very far from the Church are beginning to approach

---

1. Translated by B. Jakim, Princeton University Press, 1997.
2. *At the Watersheds* includes major treatises on reverse perspective, on Platonic idealism, and on the metaphysics of names.
3. From Lecture Seven.

[1]

her again."[4] "I have tried to clarify, first of all, the notion of the Christian world-understanding; secondly, the relation of Renaissance culture to Christian culture; and, thirdly, the logical premises of all this."[5] He affirms that there are two types of world-understanding or culture: the Renaissance world-understanding, which is anti-Christian (and even anti-religious) in nature and whose treasure lies in man; and the Medieval world-understanding, which is Christian in nature and whose treasure lies in God. He sees Renaissance culture coming to an end and being replaced by a new Middle Ages.

Florensky tries to show that science is not necessarily the handmaid of rationalistic positivism and atheism, but is compatible with Christian theology and mysticism. He sees the new epoch, the new Middle Ages, dawning in the first decades of the 20th century—it will be an epoch of the fusion of science and mystical faith which will replace the old mechanistic epoch of Renaissance rationalism.[6] An epoch of discontinuity, of abrupt leaps into reality rooted in the life of the spirit, will replace the old mechanistic world-view. This change will touch on every aspect of life and every discipline of knowledge. A revived Christianity will emerge that will find its experiential validation both in mysticism and in scientific inquiry. A new earth will arise.

The title implies that Florensky is seeking to elucidate the nature of the Christian "world-understanding." For Florensky world-understanding (in Russian: *miroponimanie*) connotes a theory or doctrine explaining the nature of the world and flow-

---

4. From Lecture Eleven.

5. From Lecture Seventeen.

6. For Florensky this refers to the era of European thought ranging from Descartes through Kant to the neo-Kantianism of the early 20th century.

ering into a full-blown culture—e.g., the Renaissance and Medieval world-understandings were embodied in the Renaissance and Medieval cultures. It is the intellectual atmosphere that dominates an epoch. Florensky tells us that "every world-understanding has a center, or treasure, of the spirit that is more ontological than are we ourselves. Our heart remains with it and begins to receive from it juices of life or death. It determines the main lines of the behavior of our reason, the main angles of our vision; that is, from a certain point of view the spiritual objects toward which we orient ourselves are the primary categories according to which our thought is organized, just as a drop has the same composition as the source from which it comes."[7] Florensky also sometimes uses the partly synonymous *mirovozrenie* (world-view) and *mirochuvstvie* (world-feeling or one's sense of the world). Three different faculties are invoked: understanding, seeing, and feeling.

One thing that becomes clear from these lectures, perhaps constituting an element of the new medieval world-understanding, is that Florensky was learned (perhaps very learned) in the occult sciences—the various theosophies and parapsychologies permeating Europe, America, and Russia at the time. There is reason to think that he had personal experience of such occult phenomena as astral bodies and auric formations, but there is no evidence that he ever successfully integrated this experience into his Orthodox Christian world-understanding.

The text translated here is the one published in the definitive edition of Florensky's works, *Sochineniya v chetyryokh tomakh*

7. From Lecture Seventeen.

(Collected Works in Four Volumes), Vol. 3(2), pp. 386–488, Moscow, 1999, ed. Igumen Andronik. Extensive use is made of Igumen Andronik's notes and commentary (pp. 531–550).

The lectures were read by Florensky from 11/24 August 1921 to 18 November/1 December 1921 at the Church of St. Sergius of Radonezh of the Moscow Upper Petrov Monastery. The author's clean copy of the lectures has not survived. The lectures have been preserved in a transcript consisting of sheets sewn into a single thick notebook; the author made many marginal notes in the notebook and inserted a number of loose sheets of textual variants and additions (in the present edition much of this added material is reproduced in the translator's footnotes). For whatever reason, not all the lectures have titles. Some of the material consists of rough notes—incomplete sentences, single-word headings, and a number of repetitions. The translator has tightened up the text by omitting some of the repetitions. All the footnotes are the translator's.

BORIS JAKIM

Lecture One

11/24 August 1921[1]

THE REMOVAL of its historical husk leads to Christianity's destruction, as in the case of Protestantism.

It does not include everything, but there is nevertheless a definite diagram of the Christian world-understanding. Evil lies in the absence of an ecclesial world-understanding.

Christianity was a doctrine not of the immortality of the soul, but of the resurrection of the dead. What we have now is a marginalization of Christ's resurrection, whereby awareness of the profound inner unity of Christ's teaching is lost.

The falseness of apologetics consists in the fact that it only indicates contradictions, without introducing new material that much more deeply and powerfully answers the demands of our spirit. Inner contradiction is a sign not of falseness, but of inner honesty. To reject something because it is contradictory is to fail to understand man's nature. Living thought is full of contradictions, as is the case with the antinomies of Christianity. Contradictoriness refers to the noumenal depths of our sense of the world.

1. The dates are sometimes given both in the old and new style, and sometimes only in the old style.

The difference between living and dead reactions to destructive effects.

A believer must have a living response to opposition. To have a living reaction is to feel the living power of creativeness in oneself. Christian truth must create from within a living stronghold for our striving toward Christ. Until attention is directed toward the development of a Christian world-understanding, apologetics will have only a palliative significance. Our task is the establishment of a Christian world-understanding.

The negation of Christianity by the spirit of the system. That which claims to answer all questions has no place in Christianity. Wherever we start, we will see that we should have started with something else as well; it is that way with all living things.

Our method is a dialectical one. Poetry, not a systematic textbook. The Apostle Paul did not write in organized sections. To organize by heading is to distort the essence of Holy Scripture.

The principle of music, which is profoundly and organically inward, is not systematic.

Leitmotifs and melodies will be repeated in different combinations and explained. Our lectures on the world-understanding will be historical; we will commune with certain historical personages and learn from them.

One's own place in history must be examined. Only a Christian relation to one's opinion will define one's relation to oneself. What is one's place in history?

The rhythm of our psychic life. Sleep takes up a third of our life.

Sleep provides much material for our self-knowledge. A great many revelations have come to people in sleep or light slumber. Sleep colors our psychic life. This happens gently, in a feminine way: when we change our convictions, we do not think of sources.

Private revelation. Its surges are fairly common. Few people are not in contact in their sleep with other worlds. The soul touches the deep, life-nourishing roots of reality.

Sleep reveals aspects of life little known to us. The life of cultural epochs is dominated either by daytime or by nighttime consciousness. The rhythm of daytime or nighttime forces occupies the same place, but the pitch differs: higher in one period, lower in another. History has days and nights. Periods of night are dominated by the mystical element, noumenal will, susceptibility, femininity. Daytime periods of history are characterized by a more active, superficial interaction with the world, phenomenal will, masculinity. The Middle Ages were a period of night; the modern age is a daytime period. We are now at the threshold of a new Middle Ages. In its depths the Christian world-understanding is medieval. In the modern period the present world-understanding is useless. The present return to the Christian world-understanding shows us that we are at the threshold of a Middle Ages.

Lecture Two

The Contemporary and the
Ecclesial World-Understanding.
Eschatological Signs

19 August 1921

MODERNITY can be reproached not only for lacking an ecclesial world-understanding but also—and chiefly—for thinking that a general world-understanding is a superfluous luxury. The Apostle Paul's words, "faith without works is dead," can serve as an example. Most people understand "works" here to mean philanthropy, while a small number include moral labor over oneself, and an even smaller number include ascetic discipline. But all three groups ignore the fact that none of this is imbued with life unless we perceive the whole world in Jesus Christ and through Jesus Christ. Otherwise all this will be something external and peripheral for us. A void will then be formed between this periphery and our heart, which will immediately be filled with evil spirits desiring to weaken our works and destroy our faith.

More and more domains of our life are leaving that of the Church and giving themselves over to the things of this world; and, finally, since man cannot live without a general world-understanding, philosophy itself begins to configure itself according to the things of this world. Nature, society, and psychology are gradually being taken away from the Church. Where can we find the domain of Christianity? Nothing is allotted to it except those crawl-spaces between worlds in which Epicurus mockingly housed the gods. This recognition of the autonomy of all the domains of life derives from the fact that the Christian world-understanding does not evolve, constituting, rather, a person's primary obligation as soon as faith is ignited in him. Instead, the Christian world-understanding that produced ecclesiastical books, national poetry, and languages, and so on—this world-understanding has gradually faded and been eclipsed by morality. For us, religion now is not liturgy, not feasts, not the idea of the world as a cosmos, not even theology—but morality only. Meantime, morality does not occupy even the smallest corner in the Christian world-understanding. Putting it crudely, it is something manufactured in Germany. The Church is made up of asceticism and mysticism; in Protestantism we get only their surrogate—morality.

Our task is to conduct a culturo-historical inquiry into the Christian world-understanding; to find out what makes for its existence, what opposes it, and what the nature of our contemporary world-understanding is. We will have to investigate the crisis of culture that began at the end of the nineteenth century.

Our contemporary events are one of the phenomena, one of the crises of destruction prior to the end of Renaissance culture.

Our principal theses are as follows: The human spirit has two aspects: nighttime and daytime, male and female. The Middle Ages and the Renaissance, sleep and wakefulness (sleep is not an absence of life, but life *sui generis*: without sleep our souls would not be nourished). Between the Middle Ages and the

Renaissance there is a period of destruction—first of the internal order, then of the social order. The Renaissance attempts, in essence, to exist outside of religion, whereas the Middle Ages lie in the religious plane, although that does not mean that men always conduct themselves in conformity with their epochs. On the contrary, conscious crimes of a particular abhorrence were possible in the Middle Ages, because the ubiquitous presence of the Church could foster profoundly evil deeds. An authentic satanism could exist precisely in the Middle Ages. But it was also possible in that period to construct one's world-understanding according to Christ. By contrast, in the Renaissance we are fated to mount a defense in the plane of the Renaissance itself, to conduct an apologetics in that very same form in which it exists at the present time; we are forced to smite culture with its own weapons.

Every destruction of culture is an eschatology, the death of an aeon, the end of an age. This produces the anxious feeling that this may be that end when all the ages and the whole world come to an end. We observe this at the present time as well. Many of the prophecies can be applied to us as well, but we know nothing about the chronology of the last times. In the Sacred History (as in all histories), the symbols are post-figurative. For example, liturgy[1] is not a simple imitation, but neither is it the figure itself. For example, the Church insistently tells us

1. The following text (abridged) appears on loose sheets appended to Florensky's notes of the Lectures: "Liturgy is symbolic. It not only symbolizes, but is the thing symbolized. The liturgical actions are not mere imitations of things that occurred in the past; nor are they the things themselves. For example, in the service celebrated on 25 December the Church tells us insistently that Christ's Nativity occurs here and now. Christ is born here and now. . . . But Christ was also born only once; this was a unique event.

that the Nativity of Christ, which took place on December 25, occurs here and now in the liturgy, at the present time; it is not a mere remembrance of an event that once took place, but something more real. It is true that Jesus Christ was born only once, that it is a unique event, but the liturgy of the feast is repeated every year. The liturgy is a representation of an event that, though it occurred at a particular moment of time, is primordially and eternally existent. This event is supratemporal, while belonging to a particular historical mloment. On feast days we begin to see another reality, whose light shines to us from beyond our empirical world.

The Eucharist is not a commemoration in the Protestant sense, but a genuine sacrifice. How should one understand it? As the possibility, given to us, of uniting with the event that exists in eternity. The end of the world, too, can produce its prefigurations in the course of the ages; just as Jesus Christ was born today as well as at a particular historical moment, the extratemporal is seen here through our everyday reality. The eschatological types of epochs have occurred repeatedly, e.g., ancient

(But since the Eternal Being participated in this event, this Being united time with eternity, enabled the temporal to participate in the eternal, and— if it be permissible to use this expression—introduced an element of temporality into eternity [*kenosis*]. And since the past is immortal for God, because for Him there is no past, present, or future, it follows that, in this sense, Jesus Christ is born eternally from the Virgin Mary.) That is why on days of feasts, when a supratemporal event is associated with a specific day, we begin to see another reality shining through our empirical existence. A feast is a window. Liturgy is the representation of an event that occurred once but that is also eternally pre-eternal. The Eucharist is not just a commemoration in the Protestant sense; it is also a genuine sacrifice, which we must think of as a possibility given to us to unite ourselves with the event that exists in eternity (the sadness experienced at the end of a feast day)."

Egypt and ancient Rome. The last epoch will be a reflection of all these types and images, and will be revealed among them; among all the other ends, there will be one when time will be no more; and just as we must fully accept that Jesus Christ was born today, so must we also take seriously every eschatological epoch; since we do not know which of these crises will be the last, each of them must be treated with full attention.

We can see a likeness of these crises in individual eschatology, in the prefiguration of our death, i.e., in sickness. We must treat with full attention every recurring instance of typhus, since we do not know which will be the last and fatal one. From the vantage point of spiritual physiology, death comes not by a fixed, invariable process, but from the coming together of a countless number of facts and from the reactions of our organism to these facts, just as a swing flies off if it is swung too hard. When a swing is in a certain position, and at the instant of its maximum deviation, a tiny addition of force suffices to make it fly off; and, conversely, a tiny subtraction of force suffices to allow it to stay on. Thus, in the life of the world there are moments when, in Shakespeare's words, "the time is out of joint";[2] and conversely, the prayer of a single righteous man can, so to speak, redeem and preserve the world. But each time one must give the matter full and serious attention, since one does not know which of these instances will be the last. But it is clear that even if we survive this time, we can expect other woes. "Oh, if only we could become free of hunger, of pestilence, etc."—this eschatological formula, repeated rather insistently (Egypt), expresses the fact that man has fallen away from Mother Earth and that humanity has grown corrupt. Hunger is a passive quarrel (with the earth)

2. *Hamlet* 1:5.

and, further, with the four elements; pestilence refers to infectious diseases in the air; then we have deluge, fire, and finally culturo-historical calamities and the degeneration of society; and then there are purely mystical calamities. The Renaissance world-understanding is the world-understanding of man who has fallen away from nature. Our epoch is attempting to gain final control over nature, raping and ravaging it, instead of listening to it; it is attempting to rationalize and enslave nature according to a predetermined plan. That is a sign of the end.

Consider Fyodorov's[3] works. They contain a complete program for the rationalization of nature and of humanity, while the religious element in his system, to which he constantly refers, is only something he has added on and which interferes with the general spirit of the system.

Signs of the crisis of western European culture have long been noticeable. For example, philosophy has self-degenerated, as the Marburg school[4] has shown with particular clarity. If philosophy has declared that it is based on an orientation toward certain propositions, and that consequently its centralization is faith, this means that we can oppose to it our own faith. Cohen has clearly asserted that philosophy is oriented toward physico-mathematical natural science, with natural science understood

---

3. Nikolai Fyodorov (1828–1903), Russian visionary whose program centered on finding technological means to modify nature and overcome death. His main idea was that man should aid God in the work of resurrection by developing technological methods to raise from the dead all the people who have ever died.

4. The Marburg school, headed by Hermann Cohen, was the center of neo-Kantianism, a dominant philosophical movement in the early 20th century.

not as science in the direct sense but as the science of a particular group of people, roughly in the style of Helmholtz. The Marburg school declares that philosophy should believe in culture. It confirms the axiom that one's faith and heart are where one's treasure is. One's heart is attuned to one's treasure. The treasure does not change in relation to the heart; the heart changes in relation to the treasure. If the treasure is assumed to lie in external culture, the heart too becomes attached to this culture. Even in the past, in the Renaissance, the heart had made a home for itself in external culture. The task people now set for themselves is the organizing of human passions. Thought was restructured by Kant and finally by the Marburg school. But there is a way out that takes us beyond rationality and pure reason; it consists in the fact that philosophy is determined by an act of noumenal will. Therefore the sciences should be studied with reference to exemplary people who have already been regenerated—with reference to the lives of the saints. The Marburg school made it possible to orient oneself on a different principle, whereas previously everyone was deeply convinced that the structure of the heart does not depend on orientation, that the heart does not depend on truth, but rather that the truth depends on man's heart and mind. Rationality is beginning to be overcome from another direction as well: by Bergson and James.

The crisis of philosophy. The husk falls away, revealing something new. Bergson. James. Immanent resistance to positivism. Proclaimed is the primacy of faith, of intuition, of noumenal will in relation to *ratio*. The crisis of the modern world-view. The complexity of being. The concept of form.

Lecture Three

On the Signs of the Epoch

25 August 1921

COMMENT: the word "epoch" is usually taken to mean a period of time, but I use it in its exact sense, such as is found in astronomy: in the sense of the beginning of a count of time, of the crossing of a "mountain ridge" of time. This word is cognate with the word "era," which means the crossing of a ridge of great magnitude, a ridge that is for us more or less absolute. To the modern and late-modern periods I oppose our contemporary one, which is fundamentally characterized by the self-overcoming of the Renaissance culture. The principles destructive of this culture also allow a new culture to appear. What forces are destroying it? A change in consciousness. A particular period of time comes to an end precisely when one type of consciousness ends and another begins.

From the culturo-anthropological point of view, this phenomenon can be characterized as an organism's self-poisoning. Nighttime consciousness is not absolutely alien to daytime consciousness: even in the daytime there can be glimmers of nighttime culture, in the way that, from the bottom of a deep well, one can see stars even at midday. In daytime culture we can descend into ourselves so deeply (here we are speaking from the ascetic-psychological point of view) that we are able to see the nighttime sky, and this sky will be even more real than the day-time one. When evening comes, stars appear everywhere. In the same way, when evening comes at a particular historical moment, almost everyone begins to see phenomena of another culture, whereas previously such phenomena had been noticed only by a few individuals who "transcended themselves" (*transcende te ipsum*—St. Augustine), who transcended the plane of rationality.

The present time is a time of twilight; a change of consciousness has already occurred, but we do not notice it only because it has occurred slowly and gradually. A colossal shift of social consciousness began at the beginning of the twentieth century. Earlier, the concept of "mysticism," for example, had been considered a form of psychopathology even in philosophical circles. For this reason many people concealed their thoughts and mentioned certain books not to everyone, but only selectively. And to mention the "Church" or "Orthodoxy" was to discredit oneself completely. To be sure, even in the past many had felt the presence of a different reality. For example, consider Tyutchev's poem:

> *O my prophetic soul!*
> *O troubled heart,*
> *How you beat on the threshold*
> *Of what seems a double reality!...*

My goal is to imbue philosophical concepts with life and concreteness. But first—a digression. "The threshold of a double reality." At first this poetic image seems to be a random one. But the deeper we look into it, the more we become convinced that it has liturgical and mystical roots. The cult of the closed door is often encountered on Egyptian stelae and signifies passage into transcendent life. In Orthodox churches the royal doors separate the sanctuary from the rest of the church; for the sanctuary represents heaven, or rather it *is* heaven. It is the world on high. Passage through the royal doors is passage into another atmosphere, a spiritual one; it is a jump in spiritual potential. Habit not only does not diminish this sense, but even deepens it as familiarity increases. This moment is terrifying; it is a kind of death, just as every *transcensus* is a jump into death. But where do we find most preeminently "the threshold of a double reality"? It is the Mother of God. It is through Her that Jesus Christ came into the world, to earth; and it is through Her that we pass into heaven. That is why She is commonly compared to the heavenly door. That is why in the hymn sung during the Small Entrance[1] She is seen as the door through which the Word came into the world.

A new group of ideas is layered on top of this fundamental idea: from the Mother, the Maternal Womb, we pass to the idea of the Small Entrance as spiritual birth. It is also death, since the altar is a place of transcendence. I say all this to show how and by what method (general) concepts are abstracted from cultic processes.

Thus, we find ourselves at the threshold of a new reality. Before our eyes the husk falls off from the culture of the past and a new culture appears, just as in springtime the greening trees open their buds. I am saying that we are seeing the advent of a new historical period; this does not mean that individual people have

1. In the liturgy, the Small Entrance is the solemn procession of the clergy to the altar led by the Book of the Gospels.—Trans.

sensed this change; it means that it has taken place in all the domains and activities of culture, that it touches upon all aspects of culture. Everywhere this change introduces the same principles of a new culture.

The first to feel this change were Tolstoy and Carpenter,[2] although for their contemporaries any notion of a crisis of science sounded like a parody. More than others, these two men possessed an intense historical sensitivity and sharp-sightedness, which were perhaps instinctive. Then there were individual events that were considered scandalous, e.g., the debate over the notion of the "life force." This notion contradicted the whole spirit and system of the science of that epoch. It was a typically medieval notion, asserting the existence of forces of another order that differed fundamentally from physical forces. The existence of *vis vitalis* was recognized and defended by Borodin and Famintsyn.[3] True, they were attacked viciously, particularly by Timiriazev;[4] however, their idea was quickly assimilated, and individual attacks against it became so frequent that it soon became a universally accepted truism that organic life differs fundamentally from inorganic life.

Starting roughly in 1900, a revolution began in science. At first it was ignored; this was done not only as a polemical tactic, but also because this tendency could not easily be popularized. Even today science is in a state of ferment, every day bringing some-

2. Edward Carpenter (1844–1929), English poet and philosopher.

3. Alexander Borodin (1833–1887), celebrated composer who was also a chemist. Andrei Sergeevich Famintsyn (1835–1918), botanist and physiologist.

4. Kliment Arkad'evich Timiriazev (1843–1920), biologist, opponent of Russian anti-Darwinism.

thing new. I am referring to the objective picture of the ideas of our time.

Let us note the formal elements of the picture of the culture of our time. There had long been complaints about specialization, not just about the impossibility of studying all the sciences, but even about the impossibility of mastering a single discipline of one science. But a strange phenomenon soon made itself known: the domains of the various sciences began to merge. Many new objects and groups of phenomena appeared whose study required more than one scientific discipline. For example, the problem of creativity—in literature, painting and sculpture, music, and so on. Biology began to merge with technology. The need to encompass everything was felt, and since individual disciplines began to merge, there appeared the tendency to apply to other specialists for help, in contrast to the former, mistrustful relation between specialists, which was rooted in the cognitive character of the epoch.

Look at 18th century portraits: an extremely common feature of the portraits is the great clarity, and even sharpness, with which individual details have been executed, as if the artist saw them through spectacles stronger than any that exist in the world. In contrast, so that this does not seem one-sided, look at icons of the 14th–15th centuries: we find clarity there, but no sharpness, no dominance of details. In the icons, the spiritual essence is in the person depicted, whereas in the 18th century portraits it lies in something much more peripheral. The medieval world-view attempted to penetrate into the depths of the essence, and therefore everything on the icon is represented with extreme clarity but without exaggeration or sharpness, without the underscoring of individual features or the splitting of objects into different parts (particularities). If the parts are opposed to one another, then each of them must be approached and studied in isolation, since they are not commensurate, do not have a common measure. Different layers of being exist in sectional view as it were, so that they cannot be glued together.

This splitting of being was not accidental. According to the dictates of reason, culture had to be sundered. But suddenly a tendency to intergrowth appeared, a tendency to coordinate methods and subordinate them to higher principles. This shows that a profound reordering had taken place in culture, that the center of gravity had shifted, just as in Jules Verne's novel *From the Earth to the Moon* the travelers suddenly noticed that under their feet they had not the earth but the moon. The same sort of gradual change in the center of gravity is taking place in our own culture, and we do not see it clearly only because of the gradualness. A restoration of the parts of the world is taking place, and if a complete world-unification has not yet occurred, there is a real movement toward it.

The question might be asked: Is it a good thing that we are living through this analytical period? But who are we to judge the destiny assigned us by God? For us it is a great gift to receive for free what our forbears only yearned for. For the same reason we also find ourselves obligated to create a church culture. We have more than enough material for this; our forbears had to defend themselves in the face of adverse historical forces, whereas we, even drifting with the current, have the possibility to create a church culture. It is left for us in humility to say: "True are all Your ways, O Lord."

In connection with the former dishonest world-view and world-understanding, there existed the idea that everything could be understood, and that if there were things that were not understood, it was because they had not yet been fully investigated. There was a sense that everything was being splintered into parts, and that the parts were elementary, spiritually two-dimensional, and deprived of internal depth and spiritual meaning, so that there was not even any reason to study them: although concrete perception necessarily includes the sense of depth, these parts were too far from one other. For example, the atomistics of the modern period stated that the atoms were too small to be studied; historians placed the beginning of historical processes in a prehistorical epoch, i.e., in a epoch so remote from us that we can say nothing about it. Here is an example that will clearly, crudely, and perhaps even caricaturishly, illustrate my thought. I was once hiking with a little boy in the woods. He told me: "there are mosquitoes here." I asked him why we can't see them. "Because they're too small," he replied. Then he told me: "there are also lions here." I asked him why we can't see them. "Because they're too big."

This type of approach is characteristic of the rationalistic world-view, which attempts to rip the veil from all mysteries and to illuminate everything with electric light.

Psychologically, it is natural for people to say that everything is very simple. This is opposite to the sense that begot philosophy—the sense of wonder. To be a philosopher is always to perceive reality as something new, as something that is never boring or stale. The adventure of the spiritual life consists in the fact that everything is renewed, first in one's consciousness and then outside oneself. The one essential thing is to transform all of reality. We must die and forget everything that seemed boring and stale; and when we awaken, all will be renewed for us; it will be beautiful and eternally joyous. And to some degree this actually happened. The second part of *Faust* represents spiritual renewal after suffering. Its beginning is depicted in hues reminiscent of the sky, an approximate vision of the primordial creature, in contrast to the task of Renaissance culture—not to wonder at anything.

Not long ago science was rushing us toward a dead end, a no exit: biology was rushing toward protoplasms, chemistry toward elements, physics toward atoms. All complex processes were fragmented into parts that were not capable of causing wonder—the most boring of world-views. Not long ago most people viewed the sky as a collection of hot frying-pans or spheres; there was nothing mysterious there (although true scientists always preserve a sense of mystery); it was nothing but "the boring songs of earth,"[5] the most ordinary blast-furnace. This view is opposite to the living human consciousness. For example, according to this view, the body is not something complex and mysterious, capable of being an object of constant contemplation and deep scrutiny, but only a certain number of elements, so many pounds of bone and muscle, etc.

But here something unexpected happened: contradicting the former dogmatized scientific understanding of reality, the life presence in the organism compelled one to look at all this differently. It turned out that these simple elements are not a dead end, but an entrance into new worlds, into a different kingdom, which produces great wonder in us. In place of the former elementarity, infinite complexity was everywhere revealed. What

5. From Lermontov's poem "The Angel."

had seemed a primitive phenomenon began to produce a unified impression of something complex. Just as the faint sound of falling drops of water can grow into the roar of a waterfall if there are enough drops, these changes in the way we look at the elements of being grew into a powerful force. It became necessary to move from one problem to another, which encompassed the first, and the latter would turn out to be even more complex. Formerly, everything had been reduced to mechanics, and mechanics to Newton's law, but it turned out that...[6]

It had been thought that protoplasm was something simple, but it turned out to be very complex and divisible into cells. Moreover, a deeper investigation of their formal composition revealed that a single one of these cells, and even a single one of its nuclei (the sexual one, for example), is a carrier of a whole set of hereditary properties, physical characteristics, psychological traits, means of verbal expression, gestures, and so on. All this is infinitely more complex than had previously been thought. But, despite this complexity, everything has a definite form—a single principle (cf. the correlation and differentiation of the elements of geometrical forms). And we perceive precisely the infinitely complex, integral form, which is essentially connected with the phenomenon. In literature and music the form is a kind of reality.

We enter into that sphere of ideas which marked the end of the medieval world-understanding—Plato, Aristotle. The philosophy of the Renaissance began with the destruction of form as a real principle. This philosophy wished to nullify form, fragmenting the whole into parts, but instead it affirmed form as a

---

6. Three lines are missing from Florensky's manuscript.

reality, even (if I may be permitted to say this) as the sole reality. Form is the principle that produces the whole diversity of different aspects. The whole precedes the parts, and the parts develop out of the whole; acknowledgment of this is a principal concession to the religious world-understanding, which is now easy to justify.

For example, how can a body be resurrected? Now, from the point of view of the theory of forms, it is easier to approach the solution of this problem. If an organism is a metaphysical form and not just the periphery of our body, then the form permeates every particle; and since the form is individual, it follows that every particle, too, is necessarily individual and personal. Every body is thoroughly individual and, as the study of tissues shows, every part of the body is permeated with individuality, even such parts as, for example, the mammary glands, and the sexual organs, too, are histologically individual. This makes it easier to understand the idea of the resurrection of the body. It would seem to be a waste of time, as well as impossible, to collect all the scattered parts; but it is a different matter if they are individualized. There will be a new process: the organism will know how to select what it needs. Just as sugar is assimilated by the organism and transformed into it, while saccharine is expelled as an unnecessary element without any trace of processing, so, in the resurrection of the body, the organism can select particles imprinted with its individuality (Gregory of Nyssa).

Another example: the Church. Contemporary tendencies aver that the Church is we, the believers. But this is a Protestant opinion. The Church does not exist because we are members of her; she does not owe her existence to us. On the contrary, she is a metaphysical form, and we can be members of her or not members of her; the Church's metaphysical reality does not suffer any loss from this. Compare the Apostle Paul's' statements "We are Christ's Body, *soma tou Christou*," and "bread is the Body of Christ, *artos—to soma tou Christou*." The use of the article makes these expressions non-identical. A definite article is not placed before the predicate, since the subject is that which is subsumed under the concept of the predicate. For example, "a student is a man, *anthropos*," whereas Jesus Christ is *o anthropos*, i.e., the Man not only according to outward traits, conduct, and so on, but also according to the identity of human nature. In Him is the fullness of humanity. He is the idea of the Man, of the Man with a capital letter. The Apostle Paul's idea is that we commune with the Body of Christ insofar as we participate in it. We by ourselves are not the Body of Christ; rather, the Eucharistic Gifts are the Very Body of Christ. The Greek has an article here. The Church is the Body of Christ, and we commune with the Church. We are *ekklesia* (without the article).

Renaissance culture is marked by the idea of continuity as opposed to discontinuity. In order to be able to negate a form, it is necessary to show that everything consists of separate elements or, in other words, that if we gradually attach some elements to others, a new form will appear. But, in fact, since precisely form is reality, this addition of elements is only the condition under which the form could appear. For example, Pushkin's poem *Eugene Onegin* precedes its letters and appears in the case of a certain selection of letters, but not in the case of a random collection of letters.

The idea of discontinuity in mathematics.

Darwinism said that the addition of infinitely small elements could produce finite changes. From the new point of view, such change is hindered not by their smallness but by the idea that no gluings will yield improvement. Such are the parts of the formal moment connected with the new world-understanding.

Lecture Four

Methods of Constructing the Kantian
World-View. The Splitting of Being into Space
and Time. The Theory of Pure Lines.
The Theory of Mutations. The Antinomic
Character of Living Thought

26 August 1921

IN OUR LAST lecture we spoke about the tendency to degrade every quality into mechanical processes; we will now speak about the process of fragmentation in time.

Renaissance culture also gave attention to the other principal form of being—time. The examination of the concept of time occurred later, and the fragmentation of being in time was performed later than the fragmentation in space: evolutionism came after mechanism.

Evolutionism does have a healthy seed—genetism, according to which an essence does not unfold in some single moment, and the spiritual meaning of an object is not exhausted by some single state, but exists in the totality of its states. But evolutionism errs when it states that this genesis is made up of infinitely small additions, so small that each separately can be considered not a creative act. Hermann Cohen confirms that my words are not a distortion of the formulation of evolutionism. Regardless of one's attitude towards him, it must be acknowledged that he is the quintessence of the ideas of the new philosophical understanding and therefore must be treated as a self-revelation of culture. In Cohen's view, he must do consciously what had previously been done unconsciously. His task is to explain the appearance of every reality. For this purpose it is necessary to find something intermediate between reality and nothing. He calls this something "ichts" ("othing"), having formed this word by removing the "N" from "Nichts." He has a certain historical justification for this neologism. In the same way, Democritus obtained the word *den* from *meden* (*meden* is no way inferior to *to den*).[1]

Science desires to show how out of nothing comes something, and how out of something comes the whole fullness of being. We already find hints of this world-view in Leibniz and in German Idealism. It then passed into natural science, and evolutionism then became lord over all the objects of our knowledge. The idea of evolutionism is connected with the idea of the infinitely small, and is an attempt to show how everything comes out of nothing. It should be stated that this addition occurs by such negligible jumps that they can be considered nothing, like a differential. The differential calculus was invented by Leibniz. His monadology is based on the principle of continuity: 1) of the series of entities; 2) of the process of perception. Nature does not perform jumps; it is impossible to pass from one extreme state to another while bypassing the middle. The idea of continuity, the principle of continuity, is the soul of Kantian

1. Florensky is saying that the Atomists derived *to den* ("that which is"), the plenum of atoms, from *to meden* ("that which is not"), the emptiness of the void.

philosophy. Science attempts to create something qualitative out of what is non-qualitative.

For example, for every immediate consciousness a straight line differs qualitatively from a curved one. But the Renaissance attempts to approach those parts of them that can be considered straight, thus performing a trick and replacing one concept by another: instead of a curve, one gets a broken line. For the immediate consciousness rest is opposite to motion, while the essence of science is the fact that we can study motion only by separating it into states as if of rest, with the result being the differential equation. The latter becomes the universal instrument of mathematics, and then of all science and the whole Renaissance epoch, since, to quote Kant, every science is a science insofar as it incorporates mathematics.

What is the differential equation? Some sort of process is occurring, and we stop it and break it up into a series of instants and see it as if in sectional view. The differential equation is a general formula suitable for the sectioning and study of any process. With this method we are not concerned with the past, with what occurred earlier in time, with whether these are people or whether they are statues who suddenly started moving again after we had stopped them. Only the present is important for us here, not the past; but for many phenomena it is precisely the past that is important. For example, in the process of magnetization it is not irrelevant for us whether the latter is increasing or decreasing.

For the differential equation, what is happening outside a given point is irrelevant. One could, so to speak, cut everything away without changing anything in the differential equation. Science asserts that one can cut off a finger and examine it in isolation, in contrast to the living consciousness, for which everything is connected with everything else. The differential equation cannot capture the past of a phenomenon; it can capture only its present.

Evolutionism is a verbal expression for the differential equation and calculus. It deems important only the presence of force and the presence of state; it is fundamentally indifferent to the past, since the reality of time and the reality of space as such do not exist for it. Evolutionism denies the existence of the whole in space and time; it denies the existence of creativity, since the whole can be produced only by a creative act—suddenly, not gradually. Recent experimental investigations effectively deny the existence of all the constructs of evolutionism. Darwin himself was courageous enough to express them, but he did not draw the practical consequences from them. Afterwards, it started to become crystal clear that changes occur in species and that they occur discontinuously. A plant is planted; a new one is obtained in the image and likeness of the first. But a break occurs in the life of this plant, in this branch of the given species, which immediately yields a new species (genus). *Mutatio* is a sudden change of species in nature. It was originally understood as self-creation, but now something like a religious understanding of it has emerged.

For example, let us consider the question of the origin of man. Evolutionism denies that there is a qualitative and fundamental difference between man and animals. But if this process occurs discontinuously, if man is descended from the apes, this theory loses its anti-religious character, since a qualitative change suddenly occurs. And if man was created from the dust of the earth by a special creative act, why not then allow in principle—only in principle—that man was created by a momentary addition of spiritual qualities to the ape? Such a view is not foreign to some religious thinkers. Serapion Mashkin[2] vividly describes a herd of animals that do not differ much from human beings. One of these animals was transfigured and in its sleep saw the Light of Tabor; when it awoke it saw that it was surrounded by beasts, and it felt how alone it was. It had suddenly become a new creature. We find something similar in Feofan the Recluse.[3] Thus, the contemporary view approaches a religious understanding of things.

In the theory of mutations, the notion of pure lines is of value. The experiments of the Catholic monk Mendel in the 19th century; his opponent Timiriazev. Mendelism is of value because it alone scientifically grounds and experimentally develops a theory of heredity, affirming that there is no self-creation in nature, but only cases of the manifestation of what already exists and of what has existed, of what is mixed through conjugal links, because the joining of the opposite cannot be stable. There had been pure species—the Platonic ideas—but they were mixed together; and then these opposite properties are being separated apart.[4] Ignorant of this, Darwinism is fifty years behind contemporary science; it recognizes only gradual evolution, but

2. Archimandrite Serapion Mashkin (1855–1905), Russian theologian. Florensky's mentor.

3. Famous Russian Orthodox spiritual writer (1815–1894).

4. Marginal note: "Unnatural joinings will continue to separate apart until pure lines are formed. To speak of prehistoric species is to agree with the Bible. One can speak only of the prehistoric existence of pure species—the Platonic ideas—which were mixed together but are in the process of being purified."

today the development of being is viewed as a revolutionary current.

If we keep in mind the theory of heredity, it is easier to approach the question of the meaning of the Old Testament. Its main task was to prepare the pure flesh of the Virgin Mary for the birth of the Savior; it was entirely a process of the purification of genes. Genes are carriers of one quality or another; they are units of heredity, are transmitted by inheritance. Evolution is a process of the splitting of mixed, non-unitable genes. Heredity extends to spiritual qualities. All first-rate English and American swindlers descend from a woman who lived in the 18th century.[5] Saints of certain epochs form kindred groups. The Apostles, of the Twelve and of the Seventy, are related to one another and to Jesus Christ. There are families which, from generation to generation, produced saints, scientists, musicians. All of history is woven from a few family fabrics, which war against one another; consider the antagonism between the Montagues and the Capulets. There are families composed of saints, kings, statesmen, scientists, artists, musicians, and so on. All of history is woven from such family fabrics.

The biological task of the Old Testament was to separate out pure genes, a holy seed. If genes of sin were mixed into man's nature when he fell, in the first generation they were split into Cainites and the descendants of Abel. Just as a dirty white cloth is gradually whitened when it is washed and the dirt is removed, so humanity was cleansed in order to produce the peak of human purity, the Mother of God, while sinful branches are broken off under the weight of sin, and perish. It has been

5. Florensky does not provide any information about this woman.

remarked that families tend to gravitate toward other families with genes of similar characteristics, with genes of alcoholism, psychosis, etc., that is, one alcoholic tends to marry another alcoholic, one mentally ill person tends to marry another mentally ill person, etc. Genes are accumulated, and that which remains has been purified.[6] The task of the Old Testament was an ever-increasing refinement tending toward the delineation of the biological axis of the world; the genealogy of Jesus Christ is the central axis of history. The thread may grow thin, but the fruit is of the highest purity.

In our contemporary philosophy is being reborn an orientation that was rejected by the Renaissance. Form is emerging more and more as something that has reality; and this emergence is sudden, not part by part. A phenomenon of correlation is revealed: form cannot be changed in any of its parts without being changed in other aspects as well, even remote ones. A change in some features causes a change in other features as well.

In the theory of pure lines the process of nature is explained on the basis of prototypes. But we do not know if prototypes are given or if they are products of Divine creativity. And so the form that had seemed diffuse and ephemeral turns out to be stronger than bronze, *aere perennus*, and what had seemed stable and unshakable turns out to be weak and fragile. The iron sole erodes, but the heel, which had seemed so tender and soft, becomes harder. The storm breaks what had seemed to be firm

---

6. Marginal note: "That which is bad is cast aside. The middle becomes brighter. And so the bad genes are accumulated and destroyed. That which is more purified remains."

and stable, but the rainbow stands and is not swept away. Someone might object and say that a rainbow is not real. That is not
so. A rainbow is a kind of reality.

The more spiritual a thing is, the stronger it is; the more material a thing is, the more unstable it is. But perhaps atoms are
more stable; they are absolutely noncomplex and unchangeable,
are they not? By no means: they too decompose.[7] It would seem
that the ultimate characteristic of all things is mass, but mass too
is separated into what is no longer matter. The law of the conservation of quantity of matter has been refuted.[8] It might have
seemed that energy, or at least the correlation of energies is stable, that everything is always the same, and in the same way,
and so on; but it has turned out that the life of the world consists
not in a regular, uniform flow, but in jumps, in constant change.

Along with the law of the conservation of energy, scientists
have discovered the law of the degradation or decay of energy,
the law of entropy, the law of dissipation.[9] Together with this, a
superphysical law of spiritual accumulation has been discovered: a law of the collection of energy and the struggle of life
against death: it would seem that the universe must die, but nevertheless it grows, and consequently superphysical forces exist
which act in that manner. These are the forces of our organism,

7. Marginal note: "The form determines the parts. In the theory, forms
emerge as prototypes. Matter is weaker than tender form. The decompos-
ability of atoms is an example."

8. Marginal note: "The law of the conservation of quantity of matter no
longer exists in science; the law of the conservation of energy has not yet
been refuted, but energy degrades: the second law of thermodynamics."

9. Marginal note: "The example of physical energy, the diminution of
the quantity of matter: the world is not evolving, but going toward its
death."

or the Higher Force that leads the universe counter to the tendency to death. "And the light shineth in darkness; and the darkness comprehended it not." The darkness acts aggressively upon the light.

In other words, while remaining within the limits of the contemporary physical world-view, we must acknowledge the existence of the Cosmic Logos. Not everything is unchangeable forever. Formerly, attempts were made to prove the stability of the universe (Newton, Laplace); in the middle period this was asserted only by physics and nature-philosophy, and the 19th century arrived at the pessimistic conclusion that the end of the world was inevitable. Either we must deny the existence of life (but this is unthinkable, since its existence is proved by the indisputable fact that the heel remains intact) or we must acknowledge the existence of a Higher Being that preserves the world. Thus, we are compelled to acknowledge the idea of creativity—it does not matter whether on a small or on a large scale.[10] The idea of Providence is confirmed more and more. The science of the near future will be entirely based on the idea of creativity and Supra-natural Force, permeating and quickening all things. Everything I am saying now schematically and in context has long been ripening in various minds and has been expressed in snatches on different occasions, becoming interwoven like lace.

This leads to another idea: creativity is life, in opposition to thingness, to things. The bearer of life is the individual, and the

---

10. Marginal note: "To acknowledge creativity in a fly or a mosquito is to acknowledge the principle of creativity; and, therefore, Providence exists."

essence of the individual is personhood. The idea of personhood has always been a familiar one in everyday life, whereas the philosophical world-view has often denied it, since personality is unitary, not complex, multifarious, and separable into parts. The whole of Renaissance culture gravitated toward *res*. For example, for Descartes the soul is *res cogitans*. In the philosophy of Spinoza and of the whole modern period there is no place for personhood; everything was conceived in the framework of thingness. Only at the end of the 19th century did one see the emergence of personalism, which attempts to develop the idea of personhood. The center of creativity is the person, and the latter is conceived as connected with the universe. Contemporary anthropology thus introduces the idea of a substantial responsibility for others, not a juridical but an ontological responsibility, the idea of responsibility for sin, for something we ourselves did not do, something like original sin.

This is particularly noticeable in Kant. He is the peak of the Renaissance world-view, and from this peak one can see things that are not otherwise visible. New tones—medieval ones—are heard in him. Of particular importance is his discovery of sinfulness as a tendency to sin, as well as his doctrine of antinomy. He pointed out that there are cracks in our reason, that rationalism self-decomposes. He explained that contradictions are a sign not of the weakness but of the vitality of human thought. The antinomy of reason is the cornerstone of the explanation in the construction of dogmas. A dogma is absolute because it is conjugately contradictory: our position is most stable when the response to a direct negation is "that is exactly what I am saying," when the broadest range between yes and no is selected.

Evolutionism and the mechanistic world-view are based on the negation of space and time, on the assertion that our forms are subjective. This characterization of human reason as an illusion was highly harmful. Space was defined solely in terms of negative characteristics (Kant); it was regarded as extension without qualities, as a vessel without walls. This led to the emptiness of Renaissance culture, to its negation of being. According to the fundamental principles of contemporary physics, the universe is finite.

Lecture Five

The Identity of Extreme Spiritualism and Materialism

14/1 September 1921

WHY IS IT necessary to abolish Descartes' separation between matter and spirit? The Renaissance world-understanding began with the negation of form, i.e., of the whole. Evolutionism and mechanism (= the negation of form) carry this out: 1) in relation to time; 2) in relation to space.[1]

The relation between quality and quantity (homeopathy). The current status of bacteriology. The medieval view: the tiniest creatures were the ones that infected. Nobody denies this now, since bacteria are only too tangible. Organotherapy. The theory of internal secretion. . . . The organism is a whole, wherein all the elements are connected. This is already a departure from the mechanical world-view, where a part suffers in isolation. Every pure element of the body receives certain substances from the blood and processes them, giving a certain part to the blood. The glands have two functions—absorption and secretion.

---

1. Marginal note: "The characterization of human experience as an illusion cuts us off from the universe; cf. evolutionism's teaching about space and time. The boundedness of space in geometry and physics. In philosophy, space and time consist of finite, though small atoms. Serapion Mashkin. His words."

Secretion is a particularly effective function. The amount and chemical composition of the secretions are almost negligible. But in their absence the organism becomes sick—in the absence of hormones.[2]

An approach to the humoral (bodily fluid) medieval theory of sickness. There is a small but important agent. The secretions of the sex organs are important. What is the reason for the Church's requirement of sexual purity? The new understanding sheds a new light. The role of genomes in the organism. Conjugal relations constitute mutual nourishment and exchange. Thus, if one's life is disorderly, then one's genomes are out of order; this is a poison, not only a physical one, but one that affects the soul. It is a poison for spiritual and for psychic life, and for the whole physical organization.

Nourishment: vulgar chemistry prescribes so many proteins, so many fats, and so on, but what they are is a matter of indifference. But now it has been discovered that there is something else, some other composition. The eastern sickness beriberi comes from nourishment with husked rice. The vitamin in rice husks provides necessary conditions for the life of the organism. In our part of the world, avitaminosis, scurvy, comes from nourishment with peeled white bread. Compare this to the Biblical teaching about pure and impure food, to the practice of fasting, with its periodicities (e.g., Wednesdays and Fridays, certain parts of the year, etc.), and to questions of church cookery.

These things, which formerly had been regarded as means for

2. The manuscript leaves out several words in this paragraph, making it somewhat garbled.

subjugating the will, are now, in the light of biological chemistry, viewed differently.

The epiphysis and the hypophysis—two glands in the brain, above and below the brain. In mysticism we have the third eye of wisdom, the hole of Brahma in Hindu sages: the mystical going out of oneself through it. In animals (lizards) it receives thermal rays. In man the epiphysis closes only gradually and has a single function: the internal secretion gland produces a special hormone; there is a connection with the manifestation of spiritual life. The hypophysis is related to maturity and sex. The hypophysis develops at puberty.

Thus, the epiphysis is connected with the higher spiritual life. "Except ye be converted, and become as little children";[3] this regeneration, this entering into the organism of a child, this overcoming of sex, is the condition for higher spiritual attainments. Thus, if we are to commune with other worlds, our functions need to be reorganized. This is possible from the present-day point of view: noumenal (internal) will can change the natural epiphysis and serve as the basis of a childlike psychology.

Religious symbolism has many foundations. Why is the symbol of mother and infant the most important symbol in Christianity? The image of the mother feeding her infant has entered into the basic iconography of the Mother of God. A woman's internal secretion has two components: 1) an erotic component, and 2) a component that liberates one from sex. The second is connected with pregnancy and breast-feeding. This indicates a higher spiritual state. The peacefulness and inwardness of such a woman.

3. Matthew 18:3.

The miracles of God's Providence never lack internal motivation. Moses's work at the exodus consisted in increasing the potential of holiness. The raising of spirit. The state of the apostle consists not in possession by external powers but in rising up to what he could be. The raising of his personhood. No rational basis exists for rejecting the Church in the mystery of the birth from the Virgin. There are indications that the greater the genius of a child, the greater the influence of the mother. For the sake of argument it could be asserted that if there were no father, the child's genius would be absolute. Conception without seed in the lower animals. Stimuli of a chemical or mechanical order. Thus, in a woman's nature there is something that enables conception without a man. A higher-order stimulus is needed. The more spiritual a being is, the more significant is the higher spiritual impetus. The descent of the Holy Spirit can serve for the conception of the Son of God.

Parthenogenesis always involves the birth of entities of male sex. In connection with this, the theory of the origin of sex: the sex is connected with the number of chromosomes. Pythagoras says: even is female, odd is male; in the present view, if the number of chromosomes is even, it is female, and vice versa. These biological data enable us to revive the ancient and medieval views. The phenomena of immunity (vaccination) and anaphylaxis. The contemporary theory of the blood. The tendency to different sicknesses ascertained on the basis of biochemical reactions.[4]

This sheds light on the religious prohibition against marriage to certain kinds of persons. Blood can be mixed in harmful ways—

4. This section has been abridged.

not only physically but also psychically, not always in the first generation, but sometimes later.

Thus, no religious practice should be rejected a priori, as insufficiently justified. That which Christianity has discovered with its illuminated eyes is being confirmed all around us and should not be rejected without proof.

So, very small anatomical details have an infinitely powerful effect on the overall state of the organism. One should not give up what has been discovered and confirmed by the ages. The principal idea of astrology regarding the connection of heaven and earth is now being confirmed...[5]

These questions are connected with the theory of heredity. Defects remain for a number of generations; there is nothing ridiculous about Biblical tales of curses extending up to the seventh generation. Biography is a new science originating in biology.

Our personal development is supratemporal; we can grasp a thing as a single whole. Take for example a Beethoven symphony: first it flows, and then it all grows together as we assimilate it. We learn to apprehend the time series of sounds as one supratemporal thing. By perceiving the law that governs them, we connect the separate elements into a single whole. Time can be viewed as the fourth coordinate of space.

5. One line is missing in the notebook.

Lecture Six

The Connectedness of Being. The Effect of Psychic Experiences on Physiology and the Effect of External States of the Body on the Soul

2 September 1921

THE RENAISSANCE WORLD-VIEW is characterized by the dualism of spirit and matter, of soul and body. For example, for Descartes matter is *res extensa* while the soul is *res cogitans*. In the Renaissance, the spirit is passive, seems to spy on things as if through a slit, not participating, but only contemplating; and matter is subordinate to mechanical laws. This world-understanding (a neo-Kantian one) has now withdrawn into the realm of legends; it has been established that there is no spirit that is incapable of acting on matter—we know an active, creative, organizing spirit. There is now no self-sufficient matter and no separate, purified spirit.

The theory of internal secretion: every tissue, every cell, etc. acts on all others, exhibits internal activity; and when blood flows into it, it processes the latter and then exudes it with a particular imprint to other parts of the body. Thus, all is connected with all, and not just with the most proximate parts of the body: every particle of the body acts upon all the others; change in one place is reflected everywhere. The body is highly interconnected. Particular cells are connected, and they are particularly important in the life of the organism; since psychic activity acts preeminently upon the whole organism, it determines the state of the whole body, in the same way that the body determines culture.

People often say "it was just something I thought," and they assume that their thought is harmless. But this small movement had an effect on the whole organism of the one thinking and it consequently provided an occasion for the rest of the organism to react to this movement and thus to interfere in the internal life of these cells. The blood river flowing through the organism is a "special kind of fluid,"[1] mysterious and of infinitely complex composition, not just something thrown together.

At the present time this is only a first attempt in the study of sciences in a new direction. In the Bible the blood is the same thing as the soul.[2] The soul lives in the blood; the blood is the body of the psychic principle, which is what imparts movement to the blood. The metaphysical composition of the blood is infinitely complex. Layered on this is the new idea of *dunamis*—the potential or power hidden beneath externalities. Phenomena can be very similar sensuously, and yet they can contain very different potentials. Sensuous identity is not equal to metaphysical identity.[3] Heredity is connected with the blood, and individually

---

1. Quotation from Goethe's *Faust* (I:1740).

2. Marginal note: "The soul is *nefesh*, the body is *basar*. Barely touching it, we immediately discover the infinite depths and mysteries of the blood river irrigating our body. 'It doth not yet appear what we shall be' (1 John 3:2)."

3. Marginal note: "With our crude sight we do not see the differences, but they exist. From the state of *dunamis* (*potentia*) heredity in the blood passes into the state of actuality (*acta*)."

different lives—spiritual and physical—are thus born. This occurs when *dunamis* becomes energy and formerly hidden powers develop.

Keeping all this in mind, it becomes easier to approach the question of sacraments. Externally, and even in external chemical composition, the materials of the sacraments do not differ in any way from ordinary materials, although they have different essences. The Holy Gifts of the Eucharist do not differ in any way in appearance from ordinary bread and wine, but metaphysically and mystically they differ in essence; the external identity does not tell us anything. The life of the body depends on internal life; the state of the body depends on psychic state. No state of the spirit is a matter of indifference for the body.

Here is an example. The physiologist Pavlov, an extreme and crude materialist, conducted experiments whose aim was to clarify the conditions under which saliva is produced. He found that it is produced prior to eating when we see food, and even when we only think about food. He showed dogs morsels of different kinds of food and noticed that the degree of saliva production differed depending on the tastiness of the morsel. It was clear that psychic state influenced digestion; and since satiety or hunger depends on the organism's assimilation of food, and the organism is subordinate to psychic states, it follows that the same food will satiate one person but not another. That is why a person of a higher spiritual life can live on almost no food at all; a saint has no problem satisfying himself with a diet on which an ordinary person could not survive. A special meaning is thus acquired by that part of the prophecy of St. Nilus the Myrrhstreamer of Mount Athos which states that, at the end of the world, the hunger will be so terrible that gold will lie untouched on the streets. But the hunger will result not from lack of food but from psychic conditions: even if they gorge themselves, the Antichrist's offspring will never be satiated because of their psychic states, whereas Christ's children will be satisfied and healthy with little.

Here is another example. Dirty thoughts are capable of changing an organism's physical and psychic state by corrupting it. This explains one of the fundamental premises of asceticism, namely that the deeper lust is embedded, the harder it is to remove it, since it has intertwined itself with the whole organism. Saints have a special physiology, which is manifested in a particular property: the sweet fragrance emanating from them and their relics. I have experienced this myself. When the relics of St. Sergius were opened, I, together with others, bent down to kiss them, and then I left the cathedral to go home. It was already night. Suddenly my attention was attracted by the fact that the air had become extremely fresh and filled with a very pleasant fragrance, the way the air is in springtime. But it was winter. There was no sign of spring, or even thawing. And so I tried to figure out what this fragrance was and where it came from. It was a fragrance of spring; it made me think of the fragrance of poplars after a storm, but there were no poplars in the vicinity and, I repeat, it was winter. Finally I figured it out: the fragrance was coming from my own lips, with which I had just touched the relics. The lives of the saints often mention the fragrance emanating from them. In contrast, medieval exorcists allude to the stink coming from demons and those possessed by them. Contemporary American psychologists have found that in states of ecstasy people emit from all the pores of their skin a fragrance similar to that of violets, and that possessed persons have a particular odor emanating from all the pores of their skin.

Spirit is manifested outwardly, in the body; it acts upon the body. The body is a symbol of spirit; it is a manifestation of spiritual states. The body is, so to speak, a spiritual state observed from outside.

Liturgy is symbolic. This is not an artificial connection but a direct expression of one thing by another: a particular action is already a corresponding state. Symbolism in liturgy is an expression of a genuine religious essence.[4]

If now we know with certainty that a male's body differs essentially from a female's body in every drop of blood, in every cell, etc.— this being the case not only crudely anatomically but also biogenetically[5]—then between a person's highest ascent and his lowest fall there lies an enormous essential difference. Despite the outward similarity of their husks, these phenomena are profoundly different.[6] Contemporary psychology has altered the whole understanding of the psychic structure of processes. Formerly, a person's psychic content was regarded as dust from various psychic states, as a bundle of associations, as a bunch of apperceptions, etc. For example, according to Herbart,[7] psychic life is a bunch of representations existing in the same chaos as Darwin's atoms. These atoms bunch together and crowd each other. Our "I," too, is only a representation together with all the

4. Marginal note: "Fragrance is, so speak, a manifestation of a spiritual state observed from outside. A symbol."
5. Marginal note: "The biochemical difference between male and female."
6. Marginal note: "A saint and a sinner are beings of different worlds. Despite the similarity of their husks, there is an internal difference between saint and sinner that will more clearly emerge subsequently, in heredity."
7. Johann Friedrich Herbart (1776–1841), German philosopher and psychologist.

others; it is something secondary, an added thought; it is not a substance but a subject of psychic states. It is our act, referring to "I" our psychic states and resembling the center point in a circle that does not contain other points lying on the periphery, but only externally unites them.[8] If we say that the soul is a substance, then it follows that it extracts its states out of itself, which is exactly what contemporary psychology says.

Earlier psychology did not have a creative principle: the organism produces everything out of itself and the mechanical law acts in things. The soul was regarded as something mechanical. American psychologists have established that we are dealing with a field of psychic states—first the whole, then separate parts. The separate parts come later. These are like folds on a separate field. They are formed by an inner creative effort of the soul. Thus, a soul has a form before it has parts. The theory of subconscious life has clarified that spiritual personality is much broader than it appears to be. It has subliminal parts which—though they are lower in nature—are rational and purposive. Our higher life—creative processes, inspiration—is governed by the supraliminal domain, while our lower life—processes connected with the life of the body—is governed by the subliminal domain. And since the body of all things is determined by the spiritual principle, every change in the field of psychic life affects all of life.[9]

How does the crystallization of psychic processes occur? Psychic life is formed extremely early, in a period about which we

8. Marginal note: "'I' is sand-like, not a unity."
9. Marginal note: "The functions of the organism are concentrated in the subliminal part. The unity and general connectedness of the soul."

remember nothing, in the first year of life, or perhaps even in the womb. This period in the process of the formation of personality is especially important. Youth is wholly personal; there is nothing empty in it; everything is full of profound meaning and interest.

An infant knows how to be astonished, to lose himself in ecstasy. He is a philosopher. There are always only a few fundamental intuitions, and everything else is crystallized around them. In childhood we see things in which another world is revealed. Thus, for example, Serapion Mashkin's first intuition, his first memory, is the sun, the image of the Light of Tabor toward which he strove his whole life. The first beneficent impressions can also be produced by one's mother: St. Augustine, Leo Tolstoy. Lermontov always remembered the blueness of his mother's eyes and the sound of her voice. That is how the personality is formed.

How does life change? It changes thanks to discontinuity in the development of every psychic process, when a certain "suddenly" occurs. Starbuck[10] tried to establish a law of religious conversion. He found, first of all, that conversions are discontinuous; even though they can occur in several stages, it is always possible to pinpoint the day and even the moment when a person touched something and it was revealed to him forever.[11] Secondly, Starbuck clarified the psychic content that exists during a conversion, e.g., when a person says "I started

10. E. D. Starbuck (1866–1947), author of *The Psychology of Religion* (1899).

11. Marginal note: "This causes a mystical state, an ecstasy. And if certain circumstances particularly strike a child, they become primary centers, roots. They can be ordinary things."

believing in God." Mystical experiences are revealed to him, illuminating what formerly had been only an abstract concept or theory. Something touches him, and suddenly a new world, unlike anything else, is revealed. He suddenly understands that this is the thing which previously he had known only theoretically. This touching of other worlds represents a discontinuous return to one's childhood: one's thoughts do not just become more sublime, they are connected precisely with one's childhood. The veil that covered one roughly at the age of seven is removed, and one returns to one's earliest impressions. To be converted is to turn around and look back. Every individual repeats in himself the history of the whole human race.[12] Haeckel's law.[13] Everyone who experiences conversion receives a childlike structure of the soul. One can say, though it is a bit of a caricature, that a genius is a grown-up child who has missed the Fall. Speaking crudely, he is Adam grown up without the Fall. Beams of Eden shine on him. Physiologically, too, genius is connected with childhood (James-Lange).

The James-Lange theory. The theory of emotions contains the great truth that there is an intimate connection between consciousness and physical states. On the one hand, the secretion of hormones; on the other, consciousness. Here the key thing is that gestures and rituals are closely connected with spiritual states. In the contemporary view, consciousness is intimately connected with physical and spiritual states. A person's garments are not a matter of indifference. Long garments—robes, vestments—prevent abrupt movements. This leads to the habit

12. Marginal note: "Each one of us contains both the fall into sin and regeneration."
13. "Ontogeny recapitulates phylogeny."

of calm, continuous gestures and movements. And since bodily movements affect the soul, if an excited person takes a calm stroll, the slow movement will inevitably calm him down, the long garments producing a calmer and slower tempo of spiritual life; and he begins to possess a greater inertia, so to speak; it is harder to discombobulate him, to knock him off his track.

This explains why Orthodoxy allows only standing or reverences down to the ground during prayer, whereas Catholics allow sitting. Let us start with a crudely material picture of a human being. A human being is principally characterized by standing on his hind feet, which perhaps led to the ability to speak.[14] What is "standing"? After all, even a chair "stands" on the floor. What is the difference between the two kinds of standing? The standing of a human being is a continuous falling in all directions and a constant effort to maintain equilibrium; it is only through exertion that we stay on our feet, since the point of support is lower than the center of gravity. Therefore in states of spiritual enfeeblement—drunkenness and so on—a human being is unable to stand. Standing is not a state but an act of the soul's continuous exertion. A standing person constantly exerts himself. "Let us stand erect"—*orthoi*; the profound implication here is that to stand erect is to be vigilant. Highly significant, therefore, is the confessor's question posed in 17th century Trebniks[15] (a question that was omitted later, when an understanding of its significance was lost): "Were you leaning against the wall during the service?" Leaning against the wall is a sin, since it is an indicator of spiritual passivity, of non-vigi-

14. Marginal note: "If a human being is an ape that stands on its hind feet. . . ."
15. The Slavonic "Book of Needs," containing prayers and services.

[51]

lance, of a particular kind of drunkenness that enables outside forces of evil to enter one's soul. Reverences down to the ground also serve this task of maintaining spiritual vigilance. Although such reverences allow us to rest completely for a moment, they then require an exertion of the entire organism, where the nervous system innervates the whole body. They reveal the human being in us: rising to our feet by an act of will, we again become human beings.

Thus, the methods of Orthodoxy teach our organism how to be human; the internal state of vigilance becomes easier and easier for us, and morally we can rise higher and higher.[16]

Both sitting and kneeling indicate an incapability of active self-determination. The Greeks considered it indecent to sit or kneel, while the Romans actually recommended sitting; here we have activity and passivity. This explains the sharp distinction between clergy and laity in Catholicism: the former act, while the latter observe; meanwhile in Orthodoxy everything belongs to the clergy. In Catholicism the lay person must be like a corpse; he must be totally in the hands of the father confessor; whereas in Orthodoxy a person can develop in his astral sphere and be born spiritually, but he must move by himself. Meanwhile, in Catholicism everything is opposite to this: everything is directed toward weakening the spirit and making a person passive. The organ, especially when a person is at rest, submerges him into in a dreamy state and facilitates the development of imagination; and imagination leads not to active attention but to

16. Marginal note: "We are nothing physically and are enfeebled spiritually. We lie in the dust and rise; we teach our organism to be vigilant, to rise higher and higher; we leave what is earthly and thrust ourselves into what is higher; we train our attention spiritually and physically."

lack of vigilance and spiritual delusion. One must do what one is commanded to do, without taking into account the internal process of life. All of a person's behavior is constrained. Psychic and external life lies under infinite constraint.

Lecture Seven

Characteristics of the Present Age.
The Theory of Functions. Neurasthenia.
Hysteria. A Theory of Spiritual Constitution.
Occult Sciences

9 September 1921

MY TASK is threefold: first, to prove the thesis that our epoch is characterized by a discontinuity or break in scientific and philosophical thought and in culture in general (the characteristic feature of this epoch being a turning back); second, to prepare the ground for the premises of the Christian world-understanding; third, to emphasize my own attitude toward freedom of thought and freedom of culture. Thought, if it is honest, inevitably arrives at the affirmation of higher spiritual values. Our contemporary thought has almost returned to its points of departure. This does not mean that these points of departure will be ecclesial in character; rather, they will be antagonistic, with an antagonism rooted, not in misunderstanding, but in a more profound and conscious resistance to Christ. Based on this characteristic, our epoch can be called eschatological. At the Last Judgment we will not be able to say, with reference to science, that we were ignorant, since this nevertheless presents a shadow of justification. I have in mind a state of eschatology where a person will be absolutely free to say "yes" either to

Christ or to the Antichrist. A time favorable for this state is approaching.[1]

The principal idea in our contemporary science is the idea of the whole. Life, not fractured into different parts, colors every individuality. A living body does not have parts, only organs. Just as a small splinter can produce pain in the whole body, so a small action directed upon the psychic life can cause a psychic and physical disturbance of the whole organism. Psychic trauma, wound of the soul, this term of contemporary psycho-pathology is a term borrowed from asceticism. "Heal the wounds of my soul" is a common expression in asceticism. Sin is viewed as a wound of the soul, and in order to restore a person's equilibrium it is necessary that heaven itself heal the wound and extract the splinter. It is clear that the soul is most subject to injury in childhood, when it has not yet acquired a protective shell; words of abuse, instances of fear, etc. can produce injuries not only in the soul but also in one's spiritual life. Therefore one can see a particularly serious and tragic meaning in the Savior's words about anyone who offends children.[2] At first glance it might appear that evil thoughts pass by without leaving any trace in the child. But these may be the most frightening words in the whole Gospel. And it is clear that the offence is not a momentary state but a preparation for psychic and perhaps spiritual ruin, in which the child himself is perhaps not to blame.

1. Marginal note: "At the time of the arrival of the Last Judgment there will be perfect freedom: there will be no obstacles to believing or not believing. A time favorable for eschatological ideas is approaching."
2. Matthew 18:6.

Every sacrament has an element of human activity. According to contemporary theory, traumas of the soul can be healed by objectification of psychic state. From the psychological side, this objectification of sin through word, this outward exposure of sin, creates favorable ground for liberation from it; and though one cannot limit oneself to this earthly aspect, it has its significance. That is why a general confession does not have the same favorable conditions from the human side as a particular one, excluding those cases when exceptional harmony and unanimity exist in the community. Repentance consists not only in a feeling of repentance but also in the verbal exposure of psychic states; and if the father confessor is not actually a witness, the conditions cannot be favorable from the earthly side. The psychic factor is considered to be the operative one here.

Pierre Janet's theory of functional neurosis. He clarified the basis of an integral life and how it becomes disorganized. Life must be internally coordinated and revolve around one center. Religious ideas are preeminently important in guiding and coordinating life. Faith in God is indispensable, since only it, as absolute, can provide a guiding activity. Otherwise, psychic illnesses and disorganization of functions will inevitably occur. The profundity and correctness of the contemporary concept of functions.[3] For example, if a person stops praying before meals,

---

3. Marginal note (abridged): "Psychotherapy. Pierre Janet.... Functional neuroses: the essence of the matter lies in the spiritual causes of certain psychic illnesses. For the psychic life to be healthy it must be integral or whole; it must have an *idée directrisse*. Janet considers that such a guiding idea must be the religious idea. Psychopathology says that faith in God is indispensable for psychic health.... The disorganization of functions has psychic causes. What is a function?"

it would appear that there are no external factors that could lead to disturbances of the function of nourishment. But the opposite turns out to be true. In the process of nourishment one can distinguish several stages. The foundation is digestion; the superstructure is the function of nourishment + the function of chewing + the function of swallowing + the process of eating. These superstructures rise higher and higher, and the higher they rise, the more psychic activity there is. The psyche (music, interesting conversations, etc.) affects the functions of nourishment. And the higher the superstructure, the greater the influence upon it of spiritual state.

Prayer assures us that when we eat we do so not like thieves, against our conscience; it assures us that when we eat, we are fulfilling God's commandment; and all these processes acquire significance, and the function of nourishment acquires unity. In contrast, when (for example) we sit at the table and argue, the lower functions will operate as they usually do, but the higher functions will become separated from them, leading to the disintegration and the separating off of the personality.

For example, the concept of marriage: every caress refers to a physiological function; on this basis a disintegration of the personality, of sex, can occur; and every instance of antagonism in a household causes a disintegration of the personality. There is a connection here with asceticism. These higher ideas lead to the idea of God. Psychological motivation of the Church's ordo.

Disintegration of the soul results in illness—neurasthenia and hysteria. They are rooted in the loss of coordination of psychic life: one loses the sense of the reality of the world; and in hysteria one loses the sense of the reality of one's person, of one's soul. The sense of empirical reality is separated from its substantial foundations. I am detached from "I." I stop being conscious of myself as "I." The sensation of oneself in hysteria: it is as if cotton is stuffed into my ears or as if I am wearing a mask. Consequently I am playing a role and want to play it as interestingly as possible, but this causes a painful agitation, both bodily and other. In neurasthenia I hear sounds and see things as in the movies: everything is phantomlike.[4] Sicknesses resulting from the disintegration of spiritual life, from the disorganization and disorder of our life, especially the sexual life. Spiritual health lies in spiritual equilibrium. There must be an internal coordination to keep the separate elements from wobbling, to correlate them. Once we start thinking that everything is conditional, such a coldness of the grave will seize us that we will stop thinking altogether.

Let us absolutize our life in the necessary functions of our life. Seven sacraments: each canonizes and illuminates a particular function in order that it become human and not just animal in character.[5] For example, the Eucharist is the idea of eating, absolute eating. All eating and nourishment are sanctified in God; they occur not according to the elements of this world.

This close resemblance to ecclesial ideas has another side, however, a negative one: treatment by psychoanalysis is, in relation

---

4. Marginal note: "The empirical 'I' is separated from the noumenal foundations of 'I'. From the loss of the sense of the spiritual principle the sensation I have is that I am an actor. I do not know who the real I is. Neurasthenia: I understand everything but everything is illusory. It is deprived of a substantial foundation; it is phantasmagoric and phantomlike."

5. Marginal note (abridged): "The result of the disordered tempo of our psychic life because of the disorders of sex. . . . One must have the sensation that everything has absolute, mystical roots. Our life must be absolutized in certain root functions."

to Confession, the same thing as the black mass in relation to the Eucharist. This often leads to the cultivation of evil principles. If ecclesial rules demand enormous experience on the part of the father confessor and strictly forbid playing the role of an elder, how much more dangerous such treatment must be in the hands of a physician who is perhaps even a nonbeliever: it often worsens the illness.

Another current of ideas in contemporary psychopathology is the theory of psychic constitution. If psychic life is something organized, it must be colored individually—not in the genetic but in the logical sense. There must be specific types of psychic constitution, similar to temperaments; each type reacts to stimuli in a particular way, reminiscent of the medieval conception of astral types in astrology. In general, we are seeing today a rebirth of the occult sciences; this does not necessarily mean that their ideas are founded, but is only a sign of the times.

Sciences that characterize manifestations of individuality in the body are highly relevant in the study of church art, with iconography being a prominent example. The noumenal pulse of individuality is revealed differently in different places. Among the most expressive places are the face and palms (these second and third faces, as they are called).[6] Hands are especially symbolic, since we cannot control our hands to the same degree as we control our face. In the 14th and 15th centuries, the golden age of iconography, hands depicted on icons have an explicitly symbolic character. These are not just hands, but bearers of a

---

6. Marginal note: "If a symbol is that which, in and through itself, manifests the Higher Principle, then hands and face are symbolic to the highest degree."

higher principle, especially in the case of the Mother of God. Her fingers are long, tapered, somewhat thin and dry, and conical in character, indicating the spirituality, sublimity, and artistic nature of Her person, while also indicating the absence of eroticism characterizing the icon-artists and the bearers of the art. This is particularly visible on the Don Icon of the Mother of God: On Her right hand Her palm and fingers are long and Her hand is thin and dry. Her left hand is different: it is full, and one senses that it is moist and trembling. In the symbolism of hands, the right receives, the left gives. Consequently, the left hand symbolizes what is taken from us, i.e., it symbolizes our passivity, or contemplativeness. In the case of the Mother of God we see two ideas or aspects. In Her activity She lacks the moist element and in a certain sense is masculine here; this is the hand of a empress, lacking the nervous element; a strong hand, ruling the Church; a dangerous hand. In contrast, Her left hand is the hand of a mother, of a being filled with inner trepidation and with a sense of mystery. There is an antinomy here. She unites birth (passivity) with virginity (an active character).

Starting from notions of occult science about methods for exteriorizing that which is internal, we can pose the question: where are the limits of our body? Even speaking materialistically, we can say that our body is not limited to the body in the strict sense of the word: our discards and emissions bear the imprint of our individuality. Even a removed fingernail has some connection with the body; and a lizard's broken-off tail keeps wriggling, since the living connection does not break off all at once, but weakens only gradually. In human beings this connection of a removed member with the body is not as noticeable owing to the centralization of the human nervous system. A new technique has been developed which involves the photographing of the auras of our body. Neurasthenia is connected with extraordinary nervous emissions from wounded places, our etheric bodies.[7] In France such wounds are treated by using incense and rotating a small red-hot blade around the body.

7. Marginal note (abridged): "The Kilner screen, invented by the director of a psychiatric clinic in London. It makes it possible to see the astral body or auric formations. These screens are used to treat certain illnesses. Hysteria is perhaps connected with damage sustained by certain places of the astral body."

Lecture Eight

Occultism. Stigmatization.
Psychic Bodies. Exteriorization of Sensitivity

15 September 1921

I PLAN TO SPEAK about a subject inadequately called "occultism," a general name that expresses nothing. I will examine its empirical aspects, saving its theory and explanation for a future date. The word *occultus* means hidden, i.e., not yet revealed. This is a negative and ambiguous definition, and therefore the most different kinds of subjects are subsumed under the concept of occultism. Luther said that the human mind resembles a drunkard sitting on top of a barrel: constantly swaying from one side to the other, it cannot find its equilibrium. This is applicable to scientific questions: either they deny the existence of one or another sphere of phenomena, or they worship them superstitiously. Thus, at first one disposed of Butlerov[1] and Crookes[2] with mockery, but then a profound change suddenly occurred in public opinion, amounting to complete, uncritical faith. Even as in the past the miracles of the Bible were confidently explained by reference to electricity, so today they are ignorantly interpreted as occult phenomena.

1. Alexander Mikhailovich Butlerov (1828–1886), a leading Russian chemist.
2. Sir William Crookes (1832–1919), British scientist and spiritualist.

A mass of the most diverse phenomena has been generalized under the concept of occult phenomena. My task is not to explain them but to show that the usual ideas about them are false. On the one hand, with the aid of physiology, and on the other hand, with the aid of psychology, and, finally, occultism, it was shown that processes of the body are guided by psychic processes; and although this psychic activity cannot be controlled by us, it is possible—with a particular kind of training— to control the processes of the body.[3] Even by means of hypnotism it is possible to affect vasomotor activity, to stop hemorrhages, and to counteract poisoning by alcohol and other toxins. The most diverse factors of spiritual life can affect the body: an improbable acceleration of the processes of organic life can be produced, e.g., the healing of wounds, the mending of broken bones.

Western Christianity has long been familiar with stigmatization—hemorrhaging through the skin and wounds developing under the influence of the intense contemplation of Christ's wounds. Even though stigmatization can be induced by hypnosis (though I do not wish to give it such a cheap explanation), there is undoubtedly a profound difference between hypnosis-induced stigmatization and the stigmatization experienced by Francis of Assisi. Stigmatization does not exist in Orthodoxy, and it exists among Catholics because they base their psychic life on imagination.[4] Psychologically, imagination—the vivid sensuous representation of certain images—creates favorable conditions for the development of stigmatization, whereas

3. Marginal note (abridged): "Subliminal consciousness. With a certain ascetic training, one can learn to control it."
4. Marginal note: "Imagination is highly developed in Catholicism— vivid sensuous representations, a tendency to dreaminess."

Orthodox asceticism requires the expulsion of these vivid sensuous representations. Even Christ's passion can be thought about only in a spiritual way. Orthodoxy does have something similar, though opposite, to stigmatization: bodily wounds produced by demons. Thus, Anthony the Great was attacked by demons and sustained wounds, stigmata produced by negative images. And since it is not possible to dream about demons, it is clear that such wounds result not from subjective dreaming, but because the representations themselves are forced upon one as something objective.

Even for extremely sensuous vision, the limit of the body lies beyond the physical body.[5] For example, light emanating from the fingers is almost a physical phenomenon. And there are many other emissions and processes of our life, and I tend to believe that thanks to our bodies we encompass the whole universe. This idea is developed by Speransky[6] when he asks how Adam and Eve were able to see that they were naked, since after the Fall their knowledge was weakened. The answer is that previously they had not been naked, since the whole world had been their body—its being had been united with theirs and so they did not see the boundaries between their body and the world. Their body had not been not naked, since it was an internal organ; and then, after the Fall, it fell out of the world as it were, leaving only thin fluid shells in the universe.

5. Marginal note (abridged): "Where is the limit of the body? . . . I think it is nowhere. We are everywhere, but in different degrees of connection with our bodily organism."

6. Mikhail Mikhailovich Speransky (1772–1839), Russian government official and mystic.

This idea corresponds to the notion in contemporary biology that the body is formed from within, not from outside.[7] Matter fills psychic and semi-material forms. Parts of these organs, not yet filled with matter, are located outside, and therefore they are very flexible and compliant. One of these organs is the light that emanates from human beings, especially from the head.

This light is an indisputable and definite phenomenon, although it is also hard to say how one sees it. Although one sees it with the sensuous eyes, it is not coordinated with the rest of the sensory situation. It can have the same gradations—of intensity and shade—as sensuous light. There can exist a whole spectrum of emissions of differing spiritual value, and its geometrical forms are always more or less uniform. We can get an idea of the light emitted by saints from halos depicted on icons.[8] The sphere filled with light and surrounding the head (see the appropriate section in my book *The Pillar and Ground of the Truth*)[9] can be represented as a luminous vessel—nimbus or gloria—on the resurrected Christ.

Religious and iconographic symbolism is always real; it always has a certain real cause; there is nothing conventional in it. And

7. Marginal note: "Noumenal will, then the formation of the body from within."
8. Marginal note: "One can see through it; it is transparent. This light can be emitted by different human beings of different nature. The spectrum, beginning with material N-rays. This light can be of different types and is represented graphically by halos: 1. the sphere around the head; 2. like a vessel surrounding the head; 3. the whole figure in an egg-shaped shell. The halos of saints are a more or less successful representation in paints of the same thing."
9. See *The Pillar and Ground of the Truth*, trans. B. Jakim (Princeton University Press, 1997), pp. 477–478.

if we do not always see this reality, it is, first of all, because of poor execution and, secondly, because of our coarseness. Liturgy is not only a symbol but also that which is symbolized.

Human beings have supra-corporeal organs that act upon the whole world. This explains the importance of the injunction of asceticism that one not invade another human being. He can be well-disposed toward us, but nonetheless we sense a profound revulsion, as if something vital and precious has been taken away from us. On the other hand, a different human being can make us feel that we have grown healthier and that our inner anxiety has abated; sitting with him even for a brief moment, we see that we have nothing to ask him about: all the questions that had tormented us have been spontaneously resolved. In contrast, it is very harmful to commune closely with people of bad spirituality, with people who are afflicted with evil passions: an infection emanates from them and we become spiritually weaker.

It is possible to act upon these peripheral but perhaps most living parts of our organism. The skin is the same thing as the nervous system. We usually associate the word "periphery" with something unimportant, but in fact it is an important thing, and this shell can be even more sensitive. Neurasthenia is a wound of the shells. Exteriorization of sensitivity is an attempt to cut off these parts of ours and to connect them with other objects.

A sensitive layer of the body up to 12 centimeters thick has been observed under deep hypnosis. There can be several such layers, and different substances can be injected into them. Wax (especially when shaped into the form of a human being) is particularly good for storing human energy. Damage to this energy causes a person to suffer. De Rochas[10] transferred human energy to a photographic plate, and damage done to the plate caused the person severe suffering.

Only the defective religious consciousness regards miracles as tricks. But miracles are actually an example of higher powers possessed by the human essence when it is sanctified. Miracles are manifestations of supra-natural powers, but in their own plane they are natural; it would have been unnatural if Jesus Christ and those who received grace from Him did not manifest a nature of a higher order. St. Augustine: for a religious person it would be astonishing if there were no miracles.

The question may be asked why miracles occur so rarely in our time. On the large, overall scale, they are not so frequent, but on the small scale everyone encounters them in his personal life. What is the character of our time? Picture a valley and high mountains. We live in the valley, in ordinary air, and only those who live in the high mountains are touched by spiritual clouds. In the evening the clouds descend and touch everyone, not only those who dwell on the mountains. And so in our epoch the "evening" has come: we notice the heightening of our mystical sensitivity on the one hand, and on the other, the descent to us of Divinity, of other worlds. Today miracles are immeasurably closer to us, as are all mystical phenomena, both gracious and dark, the nearness of the deceased and of dark powers.

For those living in the positivist lowlands, both white and black "grace"' are equally remote and alien. But when great responsi-

---

10. Eugène Auguste Albert de Rochas d'Aiglun (1837–1914) was a French parapsychologist.

bility is required of us, it is then that our sensitivity is heightened.[11]

Discussion of the auras of the organism. Besides their conventional meaning, embracing, kissing, and hand-shaking also cause an exchange of energies, an inner union with the other person. The different effects of different kinds of clothing. This explains the requirement of asceticism that one should refrain from sitting down and also from wearing the clothes of a wicked person; and, conversely, things belonging to a person full of spiritual energy have a beneficial effect. Specific pluses and minuses characterize communion with other persons. If you examine your own experience, you will see that this is true. "See then that ye walk circumspectly" (Ephesians 5:15). This explains the meaning of *starchestvo*[12] as communion: the starets not only sets a moral example and instructs his disciples, but also carries their souls to fruition in the womb of his astral body. Antiquity was familiar with this. In the *Theaetetus* Plato tells us how mere communion with Socrates affected people. His mere presence in a room heightened a person's knowledge; and, conversely, his presence soothed the anger of those who disputed with him.

11. Marginal note: "The earlier miracles are becoming closer to us and more habitual. Mystical phenomena keep getting closer. Experience is changing in two directions. A great experience is approaching. When we must answer the question, Are we with Christ or with the Antichrist, the sense of responsibility is heightened."

12. The Russian practice of eldership where the elder (the starets) imparts his spiritual wisdom to his disciples.

Lecture Nine

Inspiration. Telepathy.
Man's Transcending of the World

16 September 1921

IN MY LAST LECTURE I spoke about the possibility of invading another person's organism. This clarifies why mass psychology differs from individual psychology. The study of mass psychology made it possible to approach the question of the history of religions. Mauss, Hubert.[1]

The general psychophysical state of the masses does not provide a key to understanding religion, but it does explain the possibility of phenomena that can be called miraculous. This provides a basis for our discussion. It can explain the significance of the temple. That is why Ignatius of Antioch advises Christians to gather *pyknoteros*, more closely together. Contemporary psychophysiology has with sufficient clarity expressed the notion of the influence of external conditions and organic states on spiritual and ideational state. The gathering of people *pyknoteros* in the temple produces special powers for the battle against dark forces. Our weakness of faith at the present time is a consequence of Slavic—and, in particular, Russian—quarrelsomeness and contentiousness. We cannot be *pyknoteros*; instead, as

1. Allusion to the French scholars Marcel Mauss and Henri Hubert.

soon as we gather, centrifugal forces, forces of repulsion, appear; although, as an individual, each one of us can be very good. There must be external conditions that could facilitate the inner gathering.[2]

A few more remarks about occultism on the basis of transcendent psychology: mystical receptivity, inspiration, presentiment, creativity were all outside the field of scientific vision. When we experience inspiration, a breath from elsewhere, we receive the effects of another world, of other forces, first into the subliminal domain and then into our consciousness.[3] It has been found that a weakening of the state of our consciousness, a sleepy state, can facilitate the emergence of our inner life, of the subconscious sphere; and the latter is like a radio-telegraph system, constantly receiving messages from other entities and constantly transmitting. Studies have substantiated the validity of conjuration.

First: conjuring with mirrors. Prolonged gazing at a object such as a shiny ring or a black mirror can call forth a hypnotic state, producing visions. Pictures appear which differ from ordinary ones in that they do not conform with the surrounding reality, and are miniature. They are called pseudo-hallucinations. Calling them "hallucinations" does not necessarily mean they are imaginary or unreal. It is true that hallucinations do not correspond to ordinary perception, but ordinary perception, too, has only a symbolic character and is not an adequate image of an

2. Marginal note: "This does not give us the possibility of *being* in a church. The consciousness of the necessity of love, the beauty, the sermon, will give us, perhaps, the possibility of gathering *pyknoteros*."

3. Marginal note: "Reality is in our subconscious, and discontinuously enters our consciousness."

object, while every visual image is formed by a creative act of fantasy. In conjuration, experiences of another reality—of the past or future or perhaps even of other worlds—leap out of the subconscious.

Second: the hearing and analysis of voices in the noise of a river, a waterfall, or a seashell. "We see as in a mirror while conjuring."[4] Usually this is incorrectly taken to mean "as through dark glass," whereas it should be understood literally as "we see with the aid of a mirror while conjuring." This is an example of how many passages of the Bible and of the Fathers of the Church can be correctly understood only on the basis of certain scientific discoveries. Therefore, commentary must proceed cautiously here.

Third: the *virgula divina*, or magic rod. This is the divining rod with which specially gifted people can find sources of underground water or buried ore.

Fourth: Phenomena of agitated homes—the basis of fetishism. A fetish produced in household noise. *Treliudit'sia*—a Moscow expression for poltergeist. A planchette guided by a rational agent (I carefully say agent, not being). Telepathy—the possibility of direct spiritual communion between people in life, before death, and after death. The sending of psychic waves and impulses is connected with the appearance of something around us.[5]

How probable are the observed coincidences? Using the theory of probability, British researchers have established that such

4. This is the literal Russian Bible translation of 1 Corinthians 13:12.
5. Marginal note: "Poltergeists, apparitions. The sending of psychic impulses which evoke in us symbolic images or real representations."

phenomena are not random. The coincidence of the phenomena and deaths. These studies make a breach in the positivist world-view, and that which we read in the lives of the saints can almost be scientifically substantiated, although (I repeat) it is impossible to explain all the miraculous phenomena in the lives of the saints by these lower occult forces alone. And if one deny their validity, one must deny everything wholesale. This must be kept in mind by critics of Holy Scripture, hagiology, etc. The earlier principle of denial—Kluchevsky,[6] establishing redactions of the lives—has become a stereotype. But the cost of this attempt at debunking is now clear to us. If spiritual phenomena, miracles, exist, they must also have their laws, laws that govern the uniformity of these phenomena which are repeated in the lives of various saints.

For example, animals were not afraid of the saints. One saint crossed the Nile on a crocodile, while another—a Russian saint—fed a bear. It was not the compiler's borrowing or imagination that transformed the Nile crocodile into a Russian bear, but the fact of the unity of the saint with nature, the fact of the return of man to his royal rank in nature. These facts have never been refuted. Also, folk tales tell us of people understanding the language of animals; they tell us of the direct inner telepathic closeness and communion of people with animals. Tamers of animals act upon them not only with fear, cunning, and threats, but also through direct communion.

Consider this case. The young son of an American farmer spent all his time with animals and would tell his parents what the ani-

6. Allusion to Vasily Kluchevsky's thesis *The Lives of the Saints as a Historical Source*.

mals were feeling and what they were complaining about. He told them, for example, that the family dog was amusing itself every night by killing a neighbor's chickens. He also told them that one of the horses was complaining of a toothache, and when they examined it they indeed found a damaged tooth. The boy's fame spread; the Negroes worshipped him and brought him their animals to find out what ailed them. As the boy grew older, his abilities began to wane. This proves that animals have telepathic powers. We find the same thing among the common Russian people, who believe that holy fools can understand animals. Likewise, animals understood and obeyed saints.[7]

We are witnessing the rebirth of alchemy and astrology. Many people are perceiving a much deeper connection between man and nature than that which positivistic science desired and knew. A connection between different strata of being. Countless threads. All phenomena are reflected in every single phenomenon. A symbol is not a conventional constant but is truly connected with other phenomena. The significance of symbols in the contemporary world-understanding. Everything is in all things. One of the foundations of the medieval world-understanding is the universal connectedness and law-governedness of nature. This is now evident not only in the domains of mechanics and physics. Every sphere of being, every separate stratum of being, every phenomenon, is a microcosm, a symbol of every other phenomenon. The revelation of phenomena in a symbol, of everything in all things; but the manifestations can be particularly vivid in separate elements. The holy fathers, Goethe...

---

7. This paragraph has been abridged.

The face and the soul. The body is formed by the soul. I do not know whom I am looking at, but I know him more or less. He manifests himself in himself, in all his creative acts; but all this is a manifestation, not the soul, but only its formative activity. Everything is the soul's symbol. The body is a symbol of the soul.

I look at a face and say: this is my brother, a person close to me. Later, when I look at a photograph, I can say: this is my brother. That statement is correct from the symbolic point of view, but not from the positivistic point of view. A photograph is a manifestation of the body, and the body is a manifestation of the human soul. The activity of the soul (of the man) is manifested here too—in the photograph—although it is manifested more distantly than the body. And since I do not know a man's essence but only his activity, I am correct when I say about the photograph: this is my father; and everyday language recognizes this. Otherwise, one would not speak about a father but instead say: these are his skin, bones, muscles, etc., not he himself.

The Renaissance world-view lacks the elements that would enable it to understand religion. Religion was reduced to morality or to individual feelings, and morality became formal in character.[8] At the present time we are witnessing a return to childhood, to childlike layers of understanding, to medieval culture. The culture of the childlike structure of the spirit; childlike psychology becomes understandable. Once again the world begins to take on a mystical character. It is interesting to note

8. Marginal note (abridged): "Inner roots. The experience of the Renaissance world-understanding lacked the elements that could provide a key to understanding Christianity and religion. Religion was reduced to morality, losing its spiritual aroma."

one aspect: the baring of inner roots, of inner life. Platonism is insufficiently understood. The *gnōthi seauton*[9] of the Temple at Delphi signifies Socratic-Platonic knowledge of the human being in oneself, not knowledge of oneself as a bundle of physical states. The task of self-knowledge is to reveal in oneself one's authentic human countenance, to know the substance, not the subject (the geometrical center of positivistic phenomena). Substance is the creative, generative cause. I know that I do not know anything; all is unreal; all is not I; all is emptiness. Until it receives an ontological basis, all knowledge is illusory.[10] But in the depths of one's being one finds "Thou art": that is the sole authentic knowledge on the basis of which everything can be built—much like Descartes' "cogito ergo sum." Although a very great distance lies between Plato and Descartes, Descartes' "cogito" is not a syllogism but a profound mystical experiencing of one's reality.

In the language of the Church, if you are to know yourself, you must touch otherworldly manifestations, you must go beyond your own manifestations—*transcende te ipsum*. And then you will find in yourself the ground of the Supreme Being. You will say to the Supreme Being: "Thou art." The response will be:

9. "Know thyself."
10. Marginal note (abridged): "Platonism began with the slogan 'know thyself.' This is not knowledge in the measurement of one's powers. Knowledge of the human being in oneself and not of only a collection of certain psychic phenomena and physical functions. To reveal the authentic human being. The subject is the center of gravity of our being, whereas the substance is the creative, generative cause. I dig into myself but do not find myself. I know that I do not know anything. . . . What is revealed in the depths of our being? The inscription in the forecourt at Delphi. That is the sole authentic knowledge on which all other knowledge can be built."

"And thou, too, art" (St. Augustine). You must transcend the confusion of the world and put aside earthly cares. That is the basis of our being.

You can find something similar to this mystical feeling, though to a much weaker degree, in the following experience: if you climb to the top of a mountain or a very high belltower and survey the whole world from a distance, you will feel your own insignificance and the insignificance of everything you see. You will leave all petty things below. When you achieve this feeling of distance from the world, your true spiritual life will be liberated. When this higher spiritual principle is revealed in you, you will see with frightening clarity that all the feelings you were accustomed to explaining as the result of circumstances are your own products; you will see that we are infinitely responsible for all things, and that we ourselves are the builders of the world. We stand then before the face of the Judge. This is a weak likeness of what we will experience at our death: there will then be no place for sentimentality or for attempts to move the Truth to pity or to deceive oneself.[11] The judgment on ourselves will be pronounced by us ourselves, not by that Truth.

The snowy peak pierces the sky and stands immobile. You will never forget it if you experience God as the Truth, for only from that moment will you begin to be conscious of yourself as a human being. As long as you have not experienced this minimum of religious experience, in your consciousness you will be only a human-like animal. It is toward this that, from all the cracks, the green sprouts of the awakening consciousness of

11. Marginal note: "Here we will be face to face with Absolute Truth. And we will not move It to pity."

man's reality, of his transcendence, are growing.[12] As long as man was fused with the elements of the world, he did not criticize them, but accepted them as something given.

This is the essence of what is called the naturalism of the Renaissance world-understanding, the affirmation of oneself as a member or part of the world. This leads to the denial of sin, to the idea that everything is as good as it can be, that evil does not lurk in man but is only a misunderstanding (Tolstoyism, Rousseauism). But from the end of the 19th century the consciousness began to emerge that the world is out of order and man is out of kilter, and that therefore everything had to be redesigned, rebuilt. Different currents emerged, first fantastical in nature, and then more scientific, e.g., genetics, the improvement of the human race, rejuvenation (in America). All these things are clear symbols that what is merely given is no longer enough—what we need is reality. We have an obscure feeling that our true home is outside this world and that we are kings in relation to nature.

Another characteristic element of our time is Bolshevism: "We will show the earth a new path"; it is an effort to reorganize the world by external means. Scriabin, Fyodorov. Based on what is higher than the world, the plan is to transform the world; there is a call for eschatology; everywhere there is a desire to create a new form of existence. Germany has at the present time something similar to Bolshevism: actualism, active mysticism, the effort to act upon the world not by external means (as in Bolshe-

12. Marginal note: "The sprouts are rising from cracks in the street by virtue of the consciousness of man's reality, of his transcendence, not by virtue of fusion with God."

[77]

vism), but by spiritual means. There can be different opinions about the means and techniques of these efforts to reorganize the world, but their basic idea is clear: man is conscious of himself as higher than the ordinary jumble of the world. German actualism is based on Goethe, Schelling, and Nietzsche, who his entire life was in essence drawn only to Christ, and rejected Christianity only because it is too accepting of evil.[13]

13. Marginal note: "A thinker in relation to whom the recent past is so blameworthy."

Lecture Ten

6 October 1921

FOR THE FORM discovered in mathematics, physics, and biology one needed to find a principle manifesting it in space and time, conceivable as reality, as the creative principle (Plato's ideas). One can predict the general direction in which scientific thought will proceed, but not the exact direction. It was made clear that man is a creative center and not just an eye looking at the world through a slit; he is not a passive spectator located outside the world, but an active participant in it. Man became aware of himself as the doctor of the world, as a creative substance. Two things were revealed: first, the higher "I," about which we spoke above; and, second, the presence in me—a presence revealed through knowledge, through practical experience—of something supra-empirical, supra-sensuous.

Man of the present age turned out to be higher than the ordinary jumble of the world. He became aware in himself of the possibility of rising above the world, of walking above it, of acting upon its elements. The awareness of this possibility could be noted even in movements that were hostile to the Church, and considered themselves to be connected only with the world. When this possibility was discovered, one began to see the revival of mystical capabilities that had atrophied. And these capabilities are the basic material that constitutes the religious world-understanding. Man's estrangement from the religious world-understanding consisted in the fact that he had stopped understanding it. His estrangement from religion was punished by his loss of the ability to understand it. The religious world-view became empty and illusory. Even in the eyes of people who, in principle, said "yes" to religion, it became alien to the mind, something that was used only in important cases, not in everyday life. Religion could no longer be found in ordinary life, in conversations, letters, or diaries. Both believers and non-believers lived on the same positivistic plane, that is, on a plane not conducive to religion; and if believers still remained, it was against all odds, thanks to the mystical instinct of life. Believers hid their belief in some corner. Protestants formulated the theory that religion was something intimate and personal. It had lost the ability to manifest itself; there was no real place for it in social life. The new spiritual revolution is connected, first of all, with the discovery in oneself of a higher "I," with the consciousness of responsibility for one's fate, the recognition of those powers that enable us to be responsible, and the recognition that man is a creator in the cosmos. The need to understand culture emerged in the form of the philosophy of history. This change of theories is extremely important. Napoleon contemptuously called such historians "men of the cabinet."

But then the ideological understanding of history began to take firm root: history was considered to be determined by abstract schemata according to which life organizes itself. This theory is a characteristic expression of the spirit of Renaissance culture, according to which history is not formed by man's creative activity but is governed by abstract thought. Man is only an observer, an eye looking through a slit. In art this spirit of the time was expressed in linear perspective. According to this view, man is not a creative actor but only an eye, and even only one point of an eye that is totally immobile. The decline of culture is connected with the decline of art. Renaissance thought is not creative. The ideological theory is replaced by the opposite tendency: historical materialism, which asserts that all theories have an economic foundation, since the very existence of theories is determined by external factors—geographic, economic, racial, etc. Economic materialism began to overcome the Renaissance world-understanding, which had been the non-creative center of being. Religious man can welcome economic materialism, since it broke the pride of Renaissance man, overcame and even annihilated him in his pretension to organize history, and revealed that, though this castle in the air is collapsing, it is being replaced by something more stable: a kind of being.

According to Descartes' philosophy, man's body is not an essential part. He would have said that man is a *res animal cogitans*. This highlights the falseness of the recent notion of immortality, and therefore of resurrection, of the resurrection of Christ and of all Christianity, where the concept of resurrection is replaced by the concept of immortality. Man began to perceive his relation to the world as something accidental, not as God's gift but as a mechanism, as something impenetrable for man himself, and even for God—whence deism and atheism. Mechanics itself was developed on the basis of the presumption that spirit could be removed from the world. The entire development of the scientific world-understanding is a series of attempts to expel God from the world; that was how the whole historical process was viewed. Physics itself is rooted in the general tendency of culture. All of mechanics is physics with such a tendency.

Why was it necessary to expel God from the world? What compelled people to try to prove that He is not included in the world? One of the factors of this tendency was the pride of Renaissance man, his aspiration to be autonomous and to arrange things in such a way that nobody would interfere with the order we have established, his desire to guarantee the stability of the walls separating us from the world. The secret thought of people of this historical period was as follows: let God exist, but we too need our little corner; we do not need miracle and grace, since they disrupt the order we have established. But what is the premise behind this pride? The premise is one that has been absorbed both by the intelligentsia and by the common people—namely, manicheism, dualism, the recognition of the existence of a principle of evil practically greater than God Himself, whatever one calls it, the devil, matter, or something else. This being is the ground that guarantees for us the possibility of autonomy. What is the attitude of the intelligentsia and the common folk toward the world? Inner aversion to the flesh and to the world, inner shame. Historically, this dualism came to us through Bogomilism[1] and spread everywhere. Consider *The Kreutzer Sonata.*[2] Aversion to the flesh is combined with the pride of the flesh; the latter is considered so strong that even grace cannot transfigure it.

Meanwhile, economic materialism attempts not just to deny but to overcome all connection with the world. It asserts that we are slaves, since it considers that we do not have a ground outside the natural elements; it asserts that we are dependent on society. Economic materialism is a thing of the past.

---

1. The doctrine of a dualist religious sect that believes the world was created by the devil.

2. Tolstoy's novella, which expresses the view that sex is evil.

It was replaced by the sacral theory. For an ideologue the program creates the instrument with which he operates and which turns out to be economically useful, and this mode of operations is connected with veneration. For economic materialism the instrument is first, and the abstract concept is created afterwards. Meanwhile, the instrument is deified as something useful (e.g., the veneration of a millstone). In contrast, the sacral theory says that cult is the source of both economics and ideology. In a sense, cult is their *prius*. First we have cult, and then we have instruments and concepts. For example, the use of meat had, initially, a cultic character—the view that a certain clan was connected with a certain spiritual center (Platonic ideas). By eating meat, people were united with this mystical essence of animals. Eating of meat was communion with a certain essence uniting a certain clan. The eating of animals was connected with their veneration. Animals were eaten as friends; they were a holy thing and at a certain moment they were offered in sacrifice; this sacrifice-offering was often conceived as self-slaughter. The same thing can be said about the cultivation of grains. It originated in the veneration of the dead. A grain is a symbol of immortality, of resurrection. Blood was poured on graves, and the eating of grains grown on graves was regarded as a communion with ancestors. Only afterwards did the sacred character of eating begin to wane. First, people communed; then they partook; then they ate; and, finally, they ate like pigs.

The main scholars of the sacral theory (Frazer, Smith, Jevons, Huber, Mauss, et al.) assert that cult is the starting point of culture. Philological analysis also supports this theory. Culture is a "future participle," even as the word "nature" is the unfolding potential of cult.

All scientific—economic and suchlike—conceptions develop by a process of secularization: on the one hand we have the formation of utilitarian concepts and, on the other, we have the formation of scientific concepts. From the point of view of such an origin of philosophical concepts it is evident that positivism is right when it asserts that metaphysical concepts lack content. It is true that philosophers argue against this view by pointing to the self-sufficiency of these concepts, but they can become such only in cult; they can be fully clarified and understood only in religion. Philosophical theories differ in that some affirm this more or less consciously while others do so unconsciously, often contradicting themselves. For example, in the West the term "idea" has become useless; it can mean practically anything. In French "ideal" = "unreal."

The countenance in a portrait is the idea of the person depicted. An icon. Icons are attempts to express the highest possible artistic vision. If the church sanctuary is a place of heaven, the boundary between it and the world is guarded by real saints. The iconostasis, separating the sanctuary from the world, is a chain of angels spiritually prohibiting our entry into the place of heaven. Since their presence is not visible to all, icons are painted in remembrance, and by means of them we ascend from the images to the proto-images.

Even the Platonic terms are taken from cult. *Anamnesis*, remembrance, is a secondary sensation from touching the heavenly world. When we look at icons, we remember the spiritual proto-images that we once had. In Plato we find the pre-existence of souls. The most direct meaning of the term *anamnesis* is the opening of the senses during initiation into mysteries, and then at the slightest hint the soul is seized by agitation and understands it. Ecclesial ideas resemble Platonic ones because Plato took his terminology from cult.

In Greek books, communion, *koinonia*, signifies the link between the heavenly and the earthly. *Parousia*, the second coming, must be understood not as an external arrival but as the metaphysical revelation of spiritual essence in the sensuous world. Another image connected with Plato. The idea of the Eucharist. Its symbol is the ecclesial chalice, the center of Christian orientation, the fixed point—the altar and the chalice. The Grail. People began to ascribe miraculous powers to it; it became a world-ruling and cosmic principle. In the *Timaeus* Plato describes how the Demiurge created the world: it mixed being with nonbeing and space with ideas. This chalice has an absolute character according to which the world orients itself. And our chalice? What can we say about it without repeating Plato? What is our metaphysics of the Eucharist? How do we conceive the sacraments? As a mixture of the Real with the illusory. Of the permanent with the transitory; the appearance of Bread united with "all things were made by Him." The same thing as Plato's krater.[3] In general it should be noted that, in reading Plato, everything must be understood in the simplest and most direct sense, since his words express experiences undergone in the mysteries.

This is an attempt to clarify the presence of cultic forms in the domain of culture. In the primordial consciousness there was nothing non-cultic; there were only cultic and non-cultic (sinful) acts. Culture is based on theocratic order.

---

3. In ancient Greece, a large vase used to mix wine and water.

Correction. It is wrong to think that the sacral theory has been lost forever. This theory is connected with the medieval consciousness. Historical life has periods of secularization and periods when all of life becomes cultic. In our time one observes a hunger for the religious wholeness of life, for the organization of life according to an *ordo*. One awaits a great synthesis of all the disciplines of science; and this is something we can expect in the not-so-distant future, since all the elements are present, and all that is needed is to give them time to crystallize.

Lecture Eleven

7/20 October 1921

OUR GOAL is twofold: first, in surveying the achievements of individual disciplines, we wish to state certain propositions from which we can draw certain conclusions; and secondly, we wish to show the existence of a shift in the spiritual life of the world, a shift in the direction of a Christian world-understanding that will establish concepts extremely favorable for the Church. People who had departed very far from the Church are beginning to come back to her again.

Myth arises from cult. It is the primordial grain in cult as a reality. Even if one does not know cult, it is possible to live a somewhat religious life; but if one knows cult, it is impossible not to live a religious life.[1] Myth is a theoretical attempt to explain a given cult. Of course, the sacral theory seems to be closer to the truth but, of course, it can satisfy neither logical nor religious needs. The sacral theory says that culture comes from cult. To deduce culture anthropologically. Reality and meaning. Reality is grounded in meaning. Two activities: one directed at reality, the other at meaning.

---

1. The following text (abridged) appears at this point on loose sheets appended to Florensky's notebook of the Lectures: "According to sacral theories, culture originated in cult, in cultic act. All philosophical and scientific notions originate in cult. The same thing goes for myth. The primordial reality in religion is not dogmas or myths, but cult, i.e., a certain concrete reality. Myths and dogmas are abstractions, theories. Without this concrete element of religion our understanding of the formulas is false. The religio-mystical

[87]

In culture we find the creation of utilities and meanings. Utilities are the same as instruments. The two activities must be linked. The two can have a common ground. There are products of our cultural activity that, at the same time, are reality and meaning. We have already approached cult. Things are meanings. Let us now approach something that has detached itself from cult—art. A statue is both a thing and a meaning, but it is not fully permeated with the latter. If we crush it, we will get nothing but powder. Logic is needed if is there is to be complete unity of thing and meaning, toward cult—toward a sanctified thing.

---

content of notions is given not to abstract thought but to experience. Ritual is the most primordial thing that can be excavated by a historian, *qua* historian. The cultic fact. (This is according to the sacral theory.) More than anything else, it resembles truth. We distinguish reality from meaning, concreteness of being from consciousness. Meaning is the existent truth. We distinguish activities directed at reality from those directed at meaning. Reality is utility. They are reducible to instruments. Can these activities be non-unitable? Consciousness would then split into two. What unites them is that there exist products of our activity which, at the same time, are realities and meanings. This is cult. Works of art resemble cult. They are a union (though a somewhat external one) of thing and meaning... There must be real meanings and abstract realities; otherwise complete unity of consciousness of self would not exist. Human activity can be exhaustively represented as the combination of three elements: *Instrumenta, Notiones, Sacra*—INS. Each group can have three modifications (N-I-S, N-S-I, etc.). . . . S, N, and I are three coordinates or aspects of every cultural phenomenon, but to different degrees. . . . Every cultural phenomenon has a sacral coordinate, having some relation to the Church. . . . What is the principle that unifies the three elements? If we wish to explain culture, we must derive it from culture itself. We can explain organism on the basis of organism itself—in its embryo. And a work of art can be explained not on the basis of the materials but on the basis of the creative conception. We must seek the explanations in the depths of human personality. This is called mysticism. Mysticism is the center at which all three coordinates originate. The three are given at the same time. What is central is a certain mystical experience—the root."

Three elements form culture: *Instrumenta, Notiones, Sacra* (Instruments, Notions, Sacra)—INS.

Three theoretical groups (I—economic; N—ideological; S—religious) multiplied by 3 = 9. The point of departure for all of them is to explain cult from a part of the manifestation of culture, from what, in itself, is not culture. Culture is derived without culture; in order to explain culture it is necessary to take the stage of not-yet-cultured man, when he was not man, for culture is the content of mankind.

There are two other aspects of INS: there is no utility without inner meaning; there is no meaning without activity. In the case of a saint, every action acquires a sacral nuance.

INS cannot be conceived as independent pathways, as separate coordinates; these are three inseparable phenomena, three sides of the same being—culture. Cf. mind, feeling, and will—the three modes of psychic life. There can be no exclusivity here, but only temporary predominance. The unity of culture. Culture, therefore, must be sacral. Fundamentally, a religious character always inheres in culture.

Only the medieval conception of ecclesial wholeness, where all of culture is permeated with the sacral principle, can be convincing for us. I do not mention I and N, because that would be trite; who does not talk about them in this context? There is no phenomenon that does not have an ecclesial place. All phenomena are oriented toward ecclesiality—positively or negatively. The black mass is an example.

What principle unites these three elements? Logically, it is possible to derive culture from an element not belonging to culture. It is possible to explain culture as arising from culture in its potential, unrevealed form, from self-unitary though unrevealed cultural creativity. The embryo of culture. It must be sought in the recesses of the human personality; without it, a human being is not a human being, but what is given to him in a hidden manner. This is mysticism, the center of INS, which unifies the latter. The root of relationships.

Lecture Twelve

20 October 1921

IN MY LAST LECTURE I pointed out that culture is one, even though it has three coordinates. It can be unified under a single characteristic, since it derives from a single sense of the world. It is this characteristic that determines culture: "where your treasure is, there will your heart be also." The mystical sense is determined by a certain value. There are three types of culture: (1) one that is based on mystical life and goes toward God, (2) one that comes from God, and (3) an indeterminate mixture that oscillates upward or downward depending on the two opposite cultures. This third type existed, for example, prior to the Renaissance epoch (Leonardo da Vinci, Botticelli), when people had not yet determined where their treasure was; the heart was wavering, and two ideas were locked in conflict: (1) the idea that the being of nature is autonomous and opaque and that God is powerless in it and (2) the idea that nature is totally transparent, that it is a manifestation of Divinity, like light diffused in air. The first idea, which regards nature as an opaque object, a solid body, is fundamental for the Renaissance world-understanding. The secret goal of this idea is mechanics, the mechanistic world-understanding. It believes in the existence of the solid body.

The problem of personality. We understand the culturo-histori-cal significance of the path upward or downward in personality.[1] This is central to the Christian world-understanding. For exam-ple, let us consider the logic of Aristotle and Aquinas. It is dom-inated by the idea that what is simple is also first, i.e., first logically, ontologically, etc. Therefore, every definition must be given through *genus proximum* and *differentia specifica*, the prox-imate general. The last idea and concept is the most general. Hegel took this to its conclusion. This circle of concepts is char-acteristic for the Renaissance world-understanding.

Plato has a different view. It is very different from ours, a fact that compels us to try to subsume it under the heading of Renais-sance logic. For Plato what is simple is not necessarily meta-physically first, and vice versa. In Plato we find a set of ideas affirming that the *prius* of knowledge can be found in what is concrete to the highest degree—*the personal countenances of a phenomenon*. And if in general the conversational tone in Plato resembles our own, if he begins a conversation from our own point of view, this is only a propaedeutic device that he uses to draw the listener more comfortably into the circle of his ideas. In the same way, we find two currents in the thought of the Neopla-tonists: (1) our own and (2) a particularly vivid and intimate cur-rent whose goal is the restoration of the pagan position (Proclus). For them the determining principles of being were gods with whom one could commune—completely concrete entities with (so to speak) iconographic countenances. The most concrete thing, personality, individuality, is logically and meta-physically first. In the Middle Ages there once existed, but is

1. Marginal note: "The idea of personality, for in the latter is defined the path toward God or away from God. The problem of personality leads to problems of the concrete."

now forgotten, the particularly logical notion (formulated by Duns Scotus) of *haecceitas*, "this-ness," the combination of those features of a thing that make it itself.

Let us consider also our elementary logic. It says that of the one we can have only a representation, not a concept, that singular classes do not exist. The fundamental point of departure for every religious world-understanding is personality; cult is veneration of concrete personality.

Pascal, whose Amulet contains the program of a philosophical system, wrote in it that his God is "the God of Abraham, of Isaac, of Jacob, not of the philosophers and scholars." In other words, his God is That Person who can be a friend of Moses, for of Moses it was said that he was a friend of God. Personality is something individual and concrete to the highest degree, not an abstract principle like Tolstoy's Good, even if it is spelled with a capital letter; and it is also the Proto-source of being and thought. God the Word is not an abstract principle but the Living Person, the only Begotten Son of God, Who could be touched by our hands. He is the Person Who is much more definite, individual, and defined than any other man. He is infinitely more defined than any of us.

Our point of reference is the Eucharist. It is a combination of concrete actions, but for us the important thing is not that it exists in general but that we partake of it. Sacraments, saints, angels, rites, etc., not to mention the Person of Jesus Christ, are metaphysically first for the religious consciousness, while logically they are the most complex, because Christ contains the whole fullness of being; every man can follow Christ, because every man can see himself in Him, can distinguish in Him such traits of his own character that he cannot see in himself because they have become darkened by dirt.[2] We can do nothing without Christ: all our actions exist in some sense in Him. He contains the whole fullness of being and at the same time He is metaphysically first. He is "the stone which . . . is become the head of the corner," the cornerstone of the whole building. Thus, the image of Christ is not simple. We collide here with the Renaissance notion that that which is first is almost an emptiness, and that the more complex a thing is, the farther it is from that which is first; the complex is necessarily conceived as something composite. We must choose between the religious understanding and the Renaissance understanding from which all religious principles have been expelled.

Our age has already understood form as the metaphysical *prius* of all things, and as something complex (though not composite) and metaphysically first. We have also understood another Platonic principle, namely that that which is first precedes the parts: the complex is not composed of the simple but, rather, the complex engenders the simple. Acknowledgment of this already implies acknowledgment of the notion of *haecceitas*. Examination of three disciplines—(1) mathematical logic, (2) the logic of mathematics, and (3) the logic of history (i.e., the investigation of the laws of historical knowledge)—leads to the notion that (1) individual concepts or singular classes exist and (2) that the logically first is not simple, and vice versa.

---

2. Marginal note: "Not an abstraction or general principle. God the Word is the primordial Logos, the source of knowledge and reality, the Son Who loves the Father. He is not an abstract principle but something individual and concrete. Not simple but infinitely complex and concrete. Christ is an individual, but each one of us can see himself in Him. He contains all truth, all individuality."

Russell (English communist, prominent scientist), Grassmann, Tissan, Vivanti, Schroeder, Boole.[3] It began with Leibniz.[4]

The concepts of point, straight line, unit, etc. are all first for us, but it is erroneous to think that that is the case, because, for example, it has been shown that in order to define the concept of a straight line it is necessary to have 27 preliminary concepts, and these 27 concepts are so difficult that about them it is impossible to say that they are psychologically first; on the contrary, to understand them requires a great internal labor. Definition through *genus proximum* and *differentia specifica* is not necessary. There exist other types of definitions, not through *genus proximum*. They have long been known both in mathematics and in everyday life. For example, a child can say that he clearly understands a thing and he can do this without recourse to any logical construction. For example, if we ask him "what is sweetness?"—he'll answer more or less: "it's when you put sugar in your mouth." This answer is, logically, beneath all criticism, although it demonstrates a clear understanding. "It's when,"

---

3. List of prominent mathematicians and logicians. It is interesting that Bertrand Russell is identified as a "communist."

4. Marginal note (abridged): "Not the abstract Protestant notion of Christ, but His reality as given in the Eucharist—that is the starting point of our entire life. Jesus Christ is the individuality that contains all other individualities; all things are in Him; all of our actions and judgments, the entire fullness of the past, present, and future, are contained in Him. Taking Christ as the starting point of our thought, we must examine everything. This goes directly against Renaissance logic. In essence, contemporary thought has overcome the Renaissance world-understanding by understanding form, that which is metaphysically first, as complex, integral, concrete. These two paths have led to a single conclusion—the existence of individual concepts. The general is not first. As one of the chief laborers in this field one should mention Russell, alone and in collaboration with Whitehead."

"if, then." These formulas are a commonplace device in mathematics. One regards x and y as defined if the totality of actions by which they are defined is given. Concrete relations can be exact determinants of the first. The first is the opposite of the general.

The same thing goes for history. Auguste Comte thought that the more abstract a thing is, the more reliable it is, the straighter it is; and this is especially true for science. But history (with its root *ist* from *oida*, "I know") is knowledge par excellence, although it does not at all conform to science as it was understood in the 19th century. For the laws of history—the "general"—have no relation to history itself: the general—the chemical, the historical, etc.—is something that does not interest us at all. Both then and now, the laws of mechanics, economics, sociology, etc. are the same. History cannot change itself into sociology, and none of these laws tells us anything; none of them enables us to understand the historical process or to characterize a given epoch. For example, we have absolutely no interest in what Zwingli had in common with all other people, and even all other animals. The key thing is not knowledge of these laws, but individuality. History is not a nomographic science; it is an iconographic science of the individual, the concrete, the singular. Although today all this sounds elementary, scientific thought had long been in a quandary over this matter.

We deal with individuality in other spheres too. For example, in astronomy we are interested in a particular combination of the general. The same thing in geology: we are interested in the earth's individual history, which occurred just once and is unrepeatable. An iconographic element must exist in all knowledge. But the question arises: Since there is an infinite number of individuals, is it not possible to get lost among them? The answer lies in the concept of typology: we study there the characteristic and individual, which is regarded as a model for others, a type in the literary sense. Eugene Onegin and King Lear are instruments of knowledge; the more concrete such an instrument is for us, the more interesting it is, and it is precisely the maximum of concreteness that reveals a new reality.

These types are always irrational. It is impossible to formulate their essence; in trying to do so, one gets an extremely sententious moral. We need not an aggregate of general concepts, but a unique concept, and the term of the latter is its name. Names are the categories of historical knowledge. A type is given by its name; everything else is only an addition to this center—with it is everything, without it is nothing. If we say "Moses," there immediately emerges before us everything we know about him; and even if we do not know anything, everything easily finds its place around him. For example, saintliness cannot exist without a person; it is not reducible to concepts. Moses is a category, an enrichment, an analogy. One can say: this is Moses. And if I apply this name to one thing or another, I arrive at knowledge of this type. History is a collection of names. Religion in general and the Holy Books in particular are infinitely precious, not because of their moral significance, but as the source of all wisdom. The relation of names is a "begat"-relation, a birth-relation. This formula is the quintessence of all history. Birth-relations are, first of all, concrete, independent, individual relations: physical, spiritual, scientific, etc. Birth also signifies artistic succession; it is the most intimate and individual relation, parallel to the causal relation in the nomographic sciences. "Abraham begat Isaac"—that is the classic formula of history; everything else is only its unfolding. It can be amplified: What is

Abraham? What is Isaac? Birth. There is no history smaller than this, for there are no smaller historical phenomena and relations between them. Why is the name the true content of history?

The name is the name of a particular person. Abraham is not a general concept. What does moral statistics talk about? It talks about unchanging things, about laws that are analogous to the laws of physics. Meanwhile, what interests us in history is individuality, the eternally new, insofar as it is a manifestation not of monotonous principles but of original ones, a manifestation, first and foremost, of man, of human personhood. The birth of a new person is a new phenomenon; and the newer a phenomenon, the more interesting it is, the more sharply it is etched in the stone tablets of history. The more typical it is, the clearer it is. An other is born in a physiological person. That is the classic schema of every history, and of the Gospel history in particular. History is typology. It is interested neither in the general nor in the singular. It is typology par excellence. The metaphysics of history gives us the types of history. The Great Canon of St. Andrew of Crete is not an allegorism but an accurate understanding of history: Biblical persons become images of spiritual states. The Old Testament history becomes a likeness of the history of an individual soul. If this were an allegory, denial of the reality of the historical persons of the Old Testament would remove the foundation from under St. Andrew, so vividly did he see the types of the murderers: Cain, Lamech, and others. These concepts are infinitely stronger, richer, and more valuable that the concepts of ethics, psychology, etc. Religious man thinks by images, the true categories of religious knowledge.

Clear to us is the significance of contemporary theories (Rickert and gnoseology) in connection with the understanding of the

Person of Jesus Christ as He appears in the Gospels. He is the universal category of all human beings and events. He and His life contain all the types of spiritual states. They serve as the category of a new religious thought. He is such a category only when He is conceived as a creative principle. The life of Jesus Christ can evolve historically and even cosmically, like all world history.

Lecture Thirteen

28 October 1921

THE *Teaching of the Twelve Apostles* contrasts two paths and enumerates their characteristics. This subject has a long genealogy in literary monuments of ancient Christianity and paganism. For example, in his remembrances of Socrates, Xenophon alludes to the myth (narrated by Socrates) according to which Hercules encountered at a crossroads two women, each of whom offered a catalogue of virtues and vices. This contrast becomes particularly significant when we speak of world-understanding.

I can sin while at the same time being a man of a particular spiritual culture—a man of the Church. Hitherto we have contrasted the paths of Renaissance and Medieval thought from the historical point of view. This is especially important at the present time, when, as I have shown, history is moving from one path to another, so that we can and even must be men of the Church. This contrast is making itself known in the very premises of thought, and it is a question not of individual contradictions but of a more profound opposition between thought and the whole culture of God; and it is worse than sin. Our task is to characterize the universal-human culture and the "scientific" Renaissance culture, which gets its name not from the word "science" but from the "scientific" world-understanding, i.e., one which continues to sharpen lines that may be partly contained in science but that lead thought to conclusions not authorized by the latter. On the other hand, the ecclesial world-understanding is conceived in the womb of something more general. This womb is the world-understanding of universal-human cultures. I note its social aspect, in contrast to the particular thought of cliques. Marxism is right in a certain sense when it says with reference to bourgeois thought that the scientific world-understanding is associated with a particular class and wishes to force the rest of humanity to think as this class does. And humanity obeys because it is weak and pliant.

We will remember that abstract thought deals with the concrete—the God of Abraham, Isaac, and Jacob. Religious man says "Peter's renunciation," and for him this is a category, a type of reality. Church literature continually forms its thought through the concrete, and it is determined by other laws of thought. In abstract thought there is no order or system; the "general" is not even expressed in such a way that one could mechanically apply it to any particular case. Meanwhile, it is clear that the logic of church hymns and the psychology of their thought do not coincide with those that are presented in manuals of logic. These modes of the expression of thought remind one of poetry. In its formal aspect, ecclesial thought differs substantially from scientific thought. We observe the same thing in the folk literature of all lands and nations. In its formal aspect, ecclesial thought resembles folk thought, the thought of the masses. Fission of thought is alien to it, and it is therefore very diverse. Its individual elements, its hymns, appear to be unconnected, but as we look at them more closely, we see a great many threads linking one thing to another. They are interwoven like lace.

It becomes increasingly clear to us that a hymn is a living organism where all cells are connected and every image is a center of diverse relations to other images, where every image is a symbol (and a symbol is a reality greater than itself, for it carries the energy of other realities). Church literature and folk literature are entirely symbolic. They are not allegories of abstract thought: One can examine different cross sections of the hymns and keep making new discoveries. Church architectonics is organically woven into a whole. Multiple threads extend through it, diversely weaving together things that at first glance appear to be linked purely mechanically.

Ecclesial acts are interwoven in the same way. Consider the Small Entrance, for example. If we look at it in a crudely empirical way as a phenomenon of the senses, it is nothing more than entry though an open door. But the Small Entrance is intertwined with a whole circle of diverse ideas. These acts serve as a mass onto which is accumulated a whole series of other experiences. This entry is also a passage from the sensuous world into a metaphysical realm, and thus it signifies spiritual birth: it is a glimmer of the Incarnation and a glimmer of the Theotokos, the idea of birth, of the mother, and hence of the earth, etc.—a multiplicity of different relations.

The ecclesial world-view cannot be stretched in a rationalized way into a single thread; in trying to do so, one will rip to pieces all the other living things. When we think with our own thought, when we think fully and seriously, for ourselves and not merely in a cabinet, we think by the method of concrete-symbolic thinking, in contrast to clique thinking, which has a relation to me only insofar as I am an entity with specific social obligations. In our words, we can sunder all living connection and destroy all symbolism, but in our depths we will always have the feeling that when, in addition to thinking, we also start living and thinking in a family, not in a cabinet, our thought will become completely permeated with life. Thought can exist separately from life, in intentional opposition to the latter. Estrangement from life is a characteristic feature of Renaissance thought; in contrast, Medieval thought lets life into itself, i.e., it allows a multiplicity of phenomena, is essentially connected with life, and shares with it a diversity of connections and the emphatic presence of their forms. Synthetic thought results in concreteness. Estrangement of scientific thought from universal-human thought leads to unreality.

What is the rationale for this estrangement? Why does one need to abandon the fragrant flowers of the culture of Plato and ecclesiality? Why does one need to separate oneself from all of humanity? In order to explain this, we must return to the concept of orientation. What prompts a bunch of people to proclaim that only their thought is valuable and to become dictators of thought? When Napoleon asked Laplace how his theory related to the question of the existence of God, Laplace answered that "I have not had any need of such a hypothesis." Such an answer is typical for the group that wishes to be the aristocracy of thought. Thus, God is a hypothesis that must gradually narrow its limits and without which—at the limit—a world-understanding must be constructed.

What does it mean concretely to dispense with God in a world-understanding? It means to transform our reality into something that arises at the wave of our magic wand. It is, first, to make these realities unreal, and then, to compose them of attributes. We can then say that we are their creators. The chief task of the philosophy of the Marburg school is to show how the whole fullness of reality arises, to construct reality out of nothing. Every such attempt is an attempt to steal a few drops of water from the ocean of Divine reality, to mix them with soap, and to blow bubbles.

Consider another aspect of this: since man is connected with the world, if the whole world is proclaimed to be a product of my reason, then it follows that nothing remains but reason itself, with its structure. But as the reality of reason is weakened, all ground is ripped out from under it. If the world is meonic,[1] then man too is abolished. When God is abolished, nature is abolished; and when nature is abolished, man himself is abolished. That is the goal of western European culture. What moves it in this direction? Here we come to the notion of orientation.

Cohen is the crown of European philosophy and also a transition to what comes next. Before Cohen it was possible to believe that reason is the first principle, but he has discovered that reason exists only insofar as it is oriented toward something else, only insofar as it is determined by a non-rational act—by an act of faith in something else.

The new philosophy is oriented toward the fact of mathematical natural-science. This group of teachings exists only alongside all other teachings. In other ages, too, there were different teachings. We take a certain reality and from it we prove the reality of Reason and of its construction, or vice versa. The teaching that certain sciences must be taken as scientific fact. A solar eclipse can be understood as due to the shadow of the moon or as the sun being eaten by a dragon (the Chinese myth). People base their beliefs on feelings, on life impulses they do not understand. The intelligentsia believes in itself, in its own

1. Characterized by non-being (from Greek *me on*).

clique; it derives the structure of reason itself from the products of reason. The laws of motion of the solar system derive from the laws of mechanics; and the latter derive either from reason (and consequently they are given and reality is not created by us) or from our clique. As soon as the intelligentsia steps away from faith in life, it becomes necessary to believe in one's clique. Between self-divinization and belief in God there is no third thing, and a man of culture without God is a self-idol (the Great Canon of St. Andrew of Crete)—and this is due not to particular transgressions but to an inner logic.

Just as those sitting on an inclined surface slide down whenever they make a movement, so here too the natural tendency is to expel God, to get rid of Him, in order to be oneself. What we find around us then are not God's gifts, but something illusory. When the concreteness of form as reality is destroyed, reality will appear as nothing more than a set of attributes. But it will turn out then that even one's own form must be destroyed. And so, psychology has transformed man into a bundle of separate states. Nature and man do not exist; only a naked self-assertion remains. Man expels himself from the world, remaining only as a concrete unit; he exits the realm of being, anticipating what will happen at the Last Judgment. Here's the paradox: man wanted to form a naturalistic world-understanding, but he destroyed nature; he wanted to form a humanistic world-understanding, but he destroyed himself as man.

As if to mock us, theosophy and anthroposophy come on the scene. God is decomposed into a series of impersonal and even unreal forces. Man is like a *matryoshka* doll; he lacks a metaphysical core. The beginning and the end are in nihilism, in the void; this is the path of death. Negation of life is the death of all reality, but affirmation of being can occur only if one is oriented toward God. Christianity is the crystallization of the purest humanity, in its purest form. In order for humanity to be revealed, the Incarnation of God had to occur in man.

Christian thought is difficult to understand because it is concrete. The Renaissance world-understanding can be represented as a chain of syllogisms and separate notions. This world-understanding is not for life. It is not difficult to describe, for it is impoverished and does not answer real life-questions. It is not a world-view by which we could live. It is a world-view of the cabinets, of the magazines.

In contrast, the Medieval world-view is based on the sense of life and on that which abides in life. It is symbolic. In each of its elements and each of its points it is connected with the entire complexity of life. Each point extends a tentacle to other points. This world-view is difficult to describe, even as it is difficult to describe the functions of a living organism; wherever we begin, it is clear that that is not the beginning: functionally, the organism is circular—it closes up within itself. Religious thought, music, poetry, and especially ecclesiality are a synthesis of all the arts, the supreme human activity. A lecture necessarily distorts the nature of the religious world-understanding. It can be learned only by living together, by talking about the weather and other everyday things. It is in these everyday things that the higher goals are visible.

Lecture Fourteen

28 October / 10 November 1921

THE DIFFICULTY of understanding the general-human world-view and the falseness—the emptiness—of the "scientific" world-view. The organism is circular and closes up in itself. This refers to art and especially to religion, the synthesis of all arts and all human activities.

Lecture as a mode of expressing the religious world-understanding only distorts it. The religious world-understanding can be learned only from life, in concrete, lapidary aphorisms. Remembrance of the past: two essays, two characterizations, St. Sergius of Radonezh and Archimandrite Fyodor Bukharev.[1] St. Sergius is the beginning and the Dante of the Russian middle ages. Archimandrite Fyodor is the end of the Russian renaissance and the beginning of the new middle ages. "The Trinity St. Sergius Lavra and Russia."[2] Two principles of culture: that which is in flesh and that which is becoming incarnate, the Hypostatic Sophia and the Divine Sophia. The femininity of the Kiev period—Sophia; the masculinity of the Moscow and Petersburg periods—the Trinity. The two progenitors of the

---

1. Alexander (Archimandrite Fyodor) Bukharev (1824–1871), important theologian who was a forerunner of the great age of Russian religious thought including Solovyov, Florensky, and Bulgakov. Bukharev was famous (or notorious) for having left the priesthood.
2. Allusion to Florensky's own essay with this title (1918).

Russian nation—Cyril and Sergius. The Heavenly Virgin, Sophia, appeared to Cyril; the Life-giving Trinity, the capability of incarnation, appeared to Sergius. The Word of St. Sergius—the "Trinity," the call to unite the Russian land in the Name of the Trinity. Culture began with communal living: the Christian community, the monasteries. The Kiev Lavra, the Trinity Lavra. The center was seen not in foreigners, but in the Russian nation speaking through the Lavra.

St. Sergius is the Russian Dante, combining all the aspects of Russian culture, all the fullness of Russian life; not a theoretical thinker or writer.

Lecture Fifteen

Archimandrite Fyodor Bukharev

3/16 November 1921

THIS LECTURE is missing in Florensky's notebook. The following extract from an unidentified book about Bukharev was found among Florensky's notes: "In the Lamb of God he saw the whole fullness of the noumenal foundations of the world. If we seek his primary intuition, it will doubtless be the Lamb of God who took upon Himself the sins of the world. Outside of Christ life is impossible, not juridically, but ontologically. In Christ is freedom. Christ's law is absolute; the order of the universe is unshakable, but just as unshakable is the energy which, like miracle, establishes this order. A higher freedom is given, which, by virtue of the fact that it values these laws, makes it possible to overcome them. This is the Apostle Paul's fundamental antinomy: law and grace. Both the law and grace are from God. By leaving the priesthood, which expressed a violation of the law, a liberation from it, Archimandrite Fyodor wished to redeem, as it were, society from the sin of not understanding the laws of Jesus Christ. By his suffering he wished to participate in Christ's passion."

Lecture Sixteen

4/17 November 1921

WITH THE PASSAGE of time there is usually a waning of interest in the fundamental elements of the religious understanding. They become boring, and no words can penetrate the consciousness until a new understanding of these words is suddenly revealed. One must speak about them either philologically in a very subtle way, or philosophically, so to speak.

Concerning orientation toward Christ. Against whom did Jesus Christ dispute? Against whom did He war? Not against sinners, not even against the priests, who were all infected with Epicureanism. He polemicized against the Scribes and Pharisees. Our present understanding of the word "Pharisee" agrees neither with the Gospel text nor with its historical meaning. We currently take this word to mean a conscious dissembler, a hypocrite who deceives others. But conflict with such people would have had little general-human interest; there would have been nothing tragic about a situation in which bad people hated Jesus Christ. But the Pharisees represented the best part of the Jewish society; they were the most intelligent and the most devoted to the interests of enlightenment. This was so in the religious respect too: they wished to draw conclusions from the Law which would regulate all of life. How elevated our own life would be if we possessed even a portion of the Pharisees' indisputable virtues. However, Jesus Christ condemned, not the Sadducees, but those who were a paragon of righteousness. He condemned neither the pagans nor the sinners, but those who deserved all manner of praise. Translated to the theoretical realm, this conflict constitutes the content of Paul's epistles, the first treatises of Christian theology, the first theoretical treatment of Christian thought. For the Christian consciousness they are characterized by the opposition between Pharisaism and something else. This is the central question. On the other hand, the Pharisees' hatred of Christ is understandable and in some sense justifiable. Indeed, if I strive with all my power to fulfill the Law, what else can be demanded from me? If a sinner is condemned, that would be just, but if it is I who am condemned, I who exert myself with all my power to fulfill the Law, that is, I who in the final analysis am a righteous man, is not this condemnation an attempt to destroy the spiritual values that I serve, an attempt to destroy a holy treasure? Here it is not a question of the Pharisees's pride. The whole power of Christ's conflict with them consists in the collision of two types of righteousness: the law of works and the law of grace. If we say that the Pharisees try in all manner of ways to observe the commandments and the sabbath, who can deny the goodness of their intentions?[1]

If the commandments are given by God, who can say that it is a bad thing to make a conscientious effort to understand them in all their breadth? For Christ, too, regarded Himself as an affirmer of the Law while condemning its most zealous fulfillers. How can this be explained?

The answer...[2] Why is the commandment, the sabbath, valu-

---

1. Marginal note: "The Pharisees must have been puzzled by Christ's 'hatred'."

2. There is an omission here in the Russian text.

able? Because it is given by God. But imagine that I have forgotten God, that I have stopped seeing Him and loving Him as the Father, while obeying with all my heart His words, His commandments. They will become evil for me, even though in themselves they are good and do not cease to be good. Celebration of the sabbath will become an idol, because for man it will be only a commandment, not an influx of Divine Power. To the extent this is true, you will be enslaved to yourself. Every moral rule, and the whole set of moral rules, will then become self-sufficient precisely because I have recognized them to be such. Thus, the commandment will become something made by my own hands, and from the path of the worship of God one will diverge onto the path of idolatry, of self-worship. And the more elevated the object of such idolatry, the more dangerous it is. The purer your life, the deeper, more dangerous, and more ineradicable will be the passion of your self-worship. On the other hand, if you fall— and your fall is a total one, face into the mud, so to speak—a correct relation to yourself will become possible: there is a chance you might stop being a self-idol and open a window through which you could communicate with God.[3]

All things that do not come from God are tinsel, and the more this tinsel glitters, the more dangerous it is. The tinsel of the Pharisees was very glittery. Blinded by it, the Pharisees soon estranged themselves from God, and their hearts became insensitive to the actions of Divine grace. And so when God Himself appeared, the Pharisees, blinded by idolatrous images, did not understand this and hated Him as the destroyer of their tinsel kingdom. By contrast, the harlots and sinners did not possess

3. Marginal note: "You might be saved—i.e., you might acquire spiritual equilibrium and live in the absolute center of being."

anything they considered valuable, but relied only on God. They have no tinsel, for they walk ahead of everyone else into the Kingdom of Heaven.

The Renaissance epoch was a renaissance of Pharisaism, having (in the new language) two names: naturalism and humanism. It proclaimed the autonomy of man and of nature. Autonomy is opposed to heteronomy, where it is considered that the law, the *nomos*, of life, is given not to us, creatures, but to other entities; in *autonomia*, on the other hand, the law is given to us by us ourselves or by nature, it does not matter which. This distinction is not an opposition; in both cases the spiritual essence is the same: both are creaturely in character and become an idol in my soul. Creatures can attempt to save themselves by their laws in the domain of any activity—philanthropy, asceticism, social activity, philosophy, science, even liturgy.[4] This tendency can have many aspects, but their spiritual essence is the same: if we value the law for the sake of the law, and as a consequence place ourselves outside the influence of grace. If we do good works from the very depths of our soul, even unto the annihilation of ourselves, but do this without any relation to God while being enraptured by our works as such—this is Pharisaism.

In the same way, even if one occupies oneself with science solely for the sake of science, not for the sake of personal fame, but wholly unselfishly, but if one does this without any relation to God—this, too, is a Pharisaic aspiration to save oneself by the Law alone, as are all Renaissance aspirations whose goal is autonomous creaturely activity.

---

4. Marginal note: "If you believe in liturgy as such, it becomes an idol."

To be autonomous is not to be blasphemous; it is to annihilate one's life in God, to annihilate the direct touching of Divine energy, and to retain only a purely theoretical and abstract concept of God. It is possible to speak of God as if one were counting roubles on an abacus. In that case, He and all objects are dead. Precisely this is the absence of an ecclesial world-understanding. There is much talk about this at the present time, even too much, but all this talk is external and positivistic in character—it is one of a number of subjects of discussion, and the point of the discussion is to express one's own personal view. What we need is *Christonomia, Theonomia.* Only the transference of our heart into God can restructure our heart and give us religious experience.[5]

Fundamentally, the Law cannot contradict grace, but the farther away it is from grace, the more it falls apart. There are two types of woes: simple woe (e.g., drunkenness) and tragic woe, when a person has a passion for something that is good in a certain sense. In the first case, a person is conscious of his vice precisely as vice, whereas in the second case he departs farther and farther from the law. In its extreme, the state of Pharisaism is a spiritual delusion, where a certain state becomes an idol. At the same time it is a very close imitation of what is genuine. And once a person has entered into this circle, there is no way out, since even an errant prayer gives joy and a feeling of satisfaction, while feeding all the other feelings, pride, conceit, arrogance, etc., so that the more his soul is filled with this tinsel glitter, the greater will be his desire to pray and the more obstinate he will be in his error and convinced of his righteousness. And only a

---

5. Marginal note (abridged): "Opposite to autonomy and heteronomy are Theonomy and Christonomy."

miracle, which is what a deep fall usually is, can open his eyes and show him how far he has gone in his error. This helps to explain the aphorism of Amvrosy of Optina, which he stated as a rule for young monks: "Do not be afraid of any sin, even fornication; rather, be afraid of fasting and prayer."

Addendum to Lecture Sixteen: Answers to Particular Questions

Any listener who objects to our words may be formally correct, for our word is not creative.

The outlook for Church life in Russia. Based on historical data and on the general feeling since the first decade of this century, I have a dual attitude toward the life of the Church and of Russia. At the present time the greater part of Russian society has a profoundly unjust, cruel, and non-Christian attitude toward the common people of Russia; recently the attitude has become contemptuous, cruel, hostile. But the common people possess a deep faith and a devotion to the historically evolved order that was connected with the Church and characterized by dogmatic unity. Meanwhile, from the middle of the 17th century, since the time of Tsar Aleksei Mikhailovich, the Russian intelligentsia had done its best to expunge in the common people the ideas and religious elements natural to them.[6] And finally it succeeded. But I think

---

6. Marginal note: "I believe, nevertheless, that the foundations of the religious world-understanding are still alive in the common people. I believe in the Russian people and I believe that it has great religious tasks to fulfill, that it will create a religious culture, although it will continue to be poisoned by various toxins for a long time to come. In the Church one continues to see the incarnation of spiritual forces in a fleshly medium."

these elements have not been expunged completely. Nevertheless, the common people had nothing to oppose to this influence, since our Christianity was a medieval one. And if that is the case, it is unjust to revile a people that has great religio-cultural tasks to fulfill. That is the optimistic side of my view.

The other side, the pessimistic one, is tragic to the extreme: for a long time yet will the common people continue to absorb the poisons that had been prepared for them. Ecclesial life is a process in which ecclesial ideas become incarnate. Therefore this life has two aspects. The first is an absolutely valuable layer: dogmas, sacraments, and even (to the extent they have not been damaged) rituals, canons, and the everyday structure of Church life. The second aspect is the element (a value in its own right, not an appendix) in which the spiritual principle is made incarnate. Dominant at the present time is a hidden docetism, a denial of the incarnation of Jesus Christ. And this definition of the Church as a gathering of believers, where we are the Church, is a heresy, since it is grounded on the autonomy of the creature which wishes to be true, valuable, and holy in itself. In such a culture the Church as the self-revelation of God is irrelevant. And we calmly reconcile ourselves with the fact that our external life is our own business, having nothing to do with God.

We cannot, and do not wish to, leave the domain of culture. For six days we live on the side; on the seventh day we go to church for two or three hours and then leave again. We do not live by the Church; we only appear in it sometimes, or even rarely.[7]

---

7. Marginal note (abridged): "If we deny the Church, we fall into docetism. About Christ it is as important to say that He became incarnate as

This means that for six days we assimilate extra-ecclesial habits and modes of thought; during our brief appearance in church our thought is adjusted a little bit, but not really altered. Meanwhile, whether inside or outside the Church, it is necessary for us to think ecclesially.

In speaking about the Church, we often quote the words: "the gates of hell shall not prevail against it" (Matt. 16:18). But we forget that these words refer to Christ's Church, not to the Russian Church. This is the formal side of the matter. As far as the essence is concerned, I believe that the Russian Church will survive as a small minority and find the right road, but not without great suffering and convulsions. A great collapse of Church life must occur, a breakup into many separate currents, all of which can be heretical and non-ecclesial. It is wrong to put the blame on people. Both sides are guilty of being permeated with a particular spirit, and they cannot help it. The entire existing Russian Church is worth nothing. It belongs entirely to the non-ecclesial culture. Everyone in Russia, even people of the Church, is a positivist. The weakness of our religious consciousness can be seen in the fact that we do not notice the greatest calamities unless they are manifested in the crudest form.

An example is theosophy, anthroposophy, and similar sciences. They are based on a real human need to experience other worlds. Their method consists in the refinement of the senses,

it is to say that He is the Son of God. The denial of Church culture is a heresy that arose under the influence of Renaissance culture. All of us are to blame for this. We say: 'External life is our own business and has nothing to do with the Church.' We inevitably live by an extra-ecclesial culture: six days we live somewhere outside, and on the seventh day we come to Church. But can one just come to Church? In essence, that is impossible."

and what they assert seems very close to the truth, not unlike what the holy fathers affirmed, but in essence they espouse the most refined positivism, which is infinitely more dangerous than the crudest materialistic positivism. The patches here are very thin. In ten years, occultism will be just as accepted as hypnotism, auric analysis, and so on. It will no longer be in conflict with positivism, and it will receive unexpected confirmation: at the present time it is still possible to use spiritual teaching to disprove crude, childish materialism, but in the future this will be impossible. This represents an enormous danger for the Church. And our teaching clergy is not only failing to take any measures against it, but does not even know what is happening in the Church. There is going to be an explosion, and it is too late to avert it.

A second example: If we start discussing dogmatic questions, every man of the Church and even every representative of the Church will say: "yes, yes, I accept all this." He'll say it politely but without listening, the way a young man answers the prattling of an old woman: "yes, yes, of course." But what exactly it is he accepts and why—this he cannot explain to himself. We do not even know the decrees of the Ecumenical Councils; for example, we often hear our believers voice arguments supporting iconoclasm. Everywhere there is inner unpreparedness. Everyone thinks positivistically, and therefore we are tempted by every kind of tasteless sectarianism and pseudo-mysticism. We have a passion for religious surrogates. We resemble the owners of a treasure chest whose key has been lost. In the same way, we have lost the understanding of the ecclesial ideas contained in our liturgical and patristic literature. That is why when I was studying at the Moscow Theological Academy I once said (perhaps frivolously and a little conceitedly) that before establishing missions to foreign lands one should first establish them for students of the theological academies. We acquire faith by touching Light and Truth. Meanwhile, our apologetics has a negative character: it tries to show the falseness and internal contradictoriness of certain propositions. But that is an incorrect approach, since, first of all, religion, too, has its antinomies, i.e., its life is antinomian in character; and, secondly, assertion of the falseness of one proposition does not yet prove the truthfulness of another. The apologetics is positively harmful, since it tries to counteract antireligious tendencies with their own means; by relying on the principles of these tendencies, it indirectly affirms their legitimacy and, thus, their right to exist.

Lecture Seventeen

10 November 1921

I HAVE TRIED to clarify, first of all, the notion of the Christian world-understanding; secondly, the relation of Renaissance culture to Christian culture; and, thirdly, the logical premises of all this. In what follows we will examine the main lines of the Christian understanding. Every world-understanding has a center, or treasure, of the spirit that is more ontological than we ourselves are.[1] Our heart remains with it and begins to receive from it juices of life or death. It determines the main lines of the behavior of our reason, the main angles of our vision; that is, from a certain point of view the spiritual objects toward which we orient ourselves are the primary categories according to which our thought is organized, just as a drop has the same composition as the source from which it comes.

The question of the deduction of the categories arises. Kant, and especially Cohen, showed, by enumerating the main angles of our vision, which of them are essentially connected with our reason, and the annihilation of which is the annihilation of our treasure.

Let us consider the domain of Renaissance thought: its self-knowledge was the criticism begotten by Kant. Kant found a historical place from which one could survey all of western

1. Marginal note: "That is how it is recognized in our consciousness, although it may be false."

[120]

European Renaissance culture. He captured the aspirations and tendencies of this culture, even those that had not been fully articulated in his time and that only his successors articulated. The impoverishment of culture. Heterogeneity of the goal: moving toward a particular goal, we reach a number of secondary goals, just as when we walk on a bad road, we can pluck some good flowers—conscious and unconscious goals. In the same way, Renaissance culture has had many valuable byproducts. The direct goal of Renaissance culture is most clearly expressed by Kant. The essence of Renaissance thought is a revolt against the Church, a rebellion against God.

The French Revolution effected a new process, enthroning the "rights of man and of nature," i.e., their autonomy, but an autonomy in relation to what? Evidently, in relation to God. The secret thought here consists in the denial of the rights of God and His appearance on earth, the denial of His Church; this is Protestantism in the broad sense of the word. But it is not enough to proclaim these rights; it is necessary to fortify the revolt. This is achieved by protecting two spheres—man and nature—against attacks from God: God is expelled from all the spheres of life that we know of. The astronomical world-view had already been formulated by Kant's time, but it was necessary to show that it was impossible to think in any other way. The whole of Kantian philosophy is a fortification of these positions with the aid of the theory of knowledge. The essence of the critical method: what fundamental principles of the life of the spirit must I affirm if I am to protect the treasure of the life of the spirit which I have put on a throne and to which my heart is attached? If I am convinced of the autonomy of human life and culture, if I am convinced that the world was created without the participation of God and that man moves himself, if I believe that the world is fortified against

God, how should I conceive the structure of my spirit in order to make clear that things could not be otherwise? The world is closed. Kant's philosophy, like all of Renaissance philosophy, is based on the principle of continuity: in space there cannot be any jumps, in time there cannot be any breaks. From the point of view of this philosophy Christ's Ascension is impossible, as is the appearance of angels. Because if one acknowledges that angels can appear, one would have to acknowledge that the world is not fortified, that no scientific construction is absolute, and that any scientific law could be overcome at any moment, which would mean that no knowledge and no culture are absolute. Consequently, how should I conceive the structure of our reason so that it be known *a priori* that the appearance of angels cannot occur? Our reasoning—in the spirit of Kant—will approximately be as follows: all right, Jesus Christ ascended to heaven, but at precisely what moment did He ascend? The task of the scientific world-understanding is to keep Christ from leaving the world, to keep Him within the bounds of sensory experience. The same thing goes for the Annunciation: before appearing, the angel had to enter through the door; before that, he had to be in the garden; before that, he had to be in the street, etc. Our task is to keep capturing him by the law of the conservation of energy and indestructibility of matter, to keep tracking him using the differential equation of motion of each of his particles. To conceive him as a connection of separate particles would go counter to the principle of continuity: there can be no break in space and time. The system of formal moments of our spirit. The system of Kantian categories is the symbol of Renaissance faith without which the revolt against God cannot succeed.

The Christian world-understanding has a different system, different principles of thought. The Renaissance world-understanding does not recognize this and that, etc. It has a negative character and therefore is very impoverished with its twelve categories. By contrast, the religious world-understanding cannot be given by means of enumeration. In essence, the whole of Christian life is a category, and there is no moment in it that is not necessary. Meanwhile, one element can take on the functions of another, in the same way that this can happen in physiological and spiritual processes. The sacraments are interchangeable, as it were; one contains others.

The whole history of dogmatic movements consisted in the creation of categories. Dogmatics was not constructed as an abstract system, but grew organically out of a single point according to which we orient ourselves—the Person of the Lord Jesus Christ. This point of the application of all spiritual forces contains potentially the entire infinite abundance of thought: all the dogmas, those that have been incorporated in the Creed and those that have not been, and even those that hitherto have not been articulated by anyone, and perhaps will never be articulated, but which are potentially contained in the Gospels. All this is concentrated in a single point: Christ came in the flesh. It is concentrated in the concrete reception of Jesus Christ, our treasure.[2]

Jesus Christ, come in the flesh, is the Word of God, and He is also Man, as well as the core of universal history, the center of being: "All things were made by Him." In Him is the fullness of life and thought. All things that are outside Him are false and

---

2. On a separate sheet in the notebook: "All of Christianity is profoundly real; all of its concepts are concrete. The dogmas are not theory, but fact. All of Christianity is found in the confession of the Son of God who

illusory. On the one hand, He is one of many, a part of the world; on the other hand, He is the all, and the world is only one of the manifestations of His creative activity. Here we *immediately* hit a contradiction, an antinomy. And every living thought hits a contradiction and lives by it. The more living the thought, the more acute the contradiction. Religious thought does not blur, but simultaneously affirms, both yes and no. Every yes is the no of something else. And when this is done, man's act of faith transcends reason and is again perceived as a unified whole. Christ includes both the category of absolute spirituality and the category of flesh.[3] From the very beginning, heresies consisted in choosing one or the other side—docetism, ebionism.[4] By contrast, living thought affirms the antinomy of the categories of religious thought.

---

came in the flesh. The whole set of dogmas evolves organically from this point and is rooted in it. Besides the ones formulated at the Councils, other dogmas exist potentially in the Gospels and in the confession of the Son of God. They can be expressed by individuals or not expressed at all. One category can perform the functions of another. The Christian categories are interchangeable to some degree. These categories are concrete, as is all of Christian thought. The whole of Patrology and the Councils were a critical investigation of these categories."

3. Marginal note (abridged): "Living with us, tangible, He did not lack anything human. He has absolute value; He is the center of the world and of history. As part of the world, He is limited; as greater than the world, He is unlimited... But that is precisely where we find the law of thought. In the fundamental laws of thought—of identity and sufficient reason—we encounter contradiction, antinomy. By an act of faith the consciousness rises above thesis and antithesis and receives the one Hypostasis of Jesus Christ. Further, we affirm the categories: on the one hand, the Absolute; on the other, the creaturely, the world."

4. Marginal note: "The first heresies weakened the contradiction: ebionism in the direction of the annihilation of the Divinity, docetism in the direction of the annihilation of the humanity. Orthodoxy affirms antinomy: when antinomy of thought is rejected, we get rationalism."

When this antinomy is rejected, we get an antireligious temperament and thought; we get non-religious man, even if all his external characteristics are ecclesial. Non-religious man attempts to overcome the antinomy not by an act of faith but by the force of his spirit. All the types of world-understanding, especially the Christian world-understanding, are formed by the dogmatic method, not the critical one. In science, too, it is necessary to establish the laws of a given state, not of a particular case. In general, it is impossible to say how any phenomenon occurs. Christianity is governed by the idea of salvation. In the Christian world-view this idea is as central as the principle of the conservation of energy is in the Renaissance world-view. Every question must be examined from the point of view of this category; every proposition must be weighed soteriologically, not abstractly and metaphysically. And if the proposition destroys the idea of salvation, it is false. There is another approach to considering religious questions: the gnostic one, which dispenses with the dogma of Incarnation and examines things not in the light of the concrete perception of the Person of Jesus Christ, but abstractly, without taking into account His relation to us. Thus, for example, even though externally they often seem to agree with the opinions of the Holy Fathers, theosophy and anthroposophy are false and anti-ecclesial, for their methods and motives are false.

Consider this example of the Christian world-understanding. Ascetic saints were worthy of seeing the uncreated light and in general of experiencing revelations of another world; for example, the ascetics of Mount Athos saw the Light of Tabor. How should one interpret this? Abstractly it can be interpreted in all manner of ways. These appearances of light can be viewed as hallucinations, as objective occult phenomena of the organism, or as a heightening of sensitivity allowing one to detect subtle radiations. There can be as many theories as you like, but they will all be false because of their methodologies. The ecclesial method is as follows. We believe that they are saved because the Church has recognized these people as worthy of salvation, i.e., as close to God. And since, on the other hand, these people regarded the Light of Tabor as the apex and crowning achievement of their ascetic feat, we must therefore conclude that the ascetic who always lived with God had to be especially united with Him at this moment of his blossoming when he saw the light, and that, consequently, this light seen by him is not something created, but the Light of God. This Light directly reveals God. If ascetics are conscious of being especially close to God at such moments, that is indeed the case. If it were not the case, it would be a monstrous error. In general, the ascetic can err, but he cannot err in this case, for if he errs in regard to what constitutes the center of his spiritual life, what would his ascetic feat be worth? What would his feat be worth if he falls to the level of a crude divinization of his own states, confusing things that even sinners do not confuse and in the case of which even they do not err? Such error is impossible, for it would contradict the idea of the salvation of the saints; consequently, this is indeed the Light of God; "I see God" is a factual expression; it is truly so.

But one could dispute this by quoting: "No man hath seen God at any time."[5] God's essence is unfathomable, and if a saint says that he has seen God, this can mean one of two things: (1) he has seen God's very essence, but that would be an assertion of pantheism, of the heresy of Messalianism, for God's transcendence and holiness would be abolished; or (2) he has seen not

5. 1 John 4:12.

God, but something else. That was Barlaam's[6] reasoning. The Orthodox understanding is that one can indeed see God, but what one sees is not His essence but His energy. The energy seems not to express the essence, but it is God insofar as He is revealed to people, and the essence is God as He is in Himself. And so, subtle dogmatic concepts were developed out of the fact of the seeing of the Light. For every entity, the essence is the side turned toward itself, whereas the energy is the side turned outward; in other words, only non-being does not have energies. The theory of symbolic world-understanding teaches that every reality, having its energy, can be illuminated by the energy of another thing. One may ask "how exactly?"—but that question is incorrect, illegitimate, and mechanistic.[7] If one thinks intently about it, this turns out to be an idle and childish question. And since thought is teleological, moving toward certain fundamental goals, every idle question is therefore ill-posed, falsely posed.

6. Fourteenth-century monk who disputed Palamas' doctrine of the real existence of the uncreated light.

7. Marginal note: "If it is considered that a saint has seen not God but his own energies, that is Barlaamism, a type of delusion. Thus, it is necessary to distinguish essence and energy in God. Every reality has essence and energy. Later in our lectures we will develop a Christian anthropology, and then a symbolic world-understanding: a thing can bear the energy of another thing. How the energy of one essence is united with the energy of another, I do not know, but I can indicate the gradations."

Lecture Eighteen

The Relation between Philosophy and Science

11 November 1921

A THINKER can be likened to a traveler who, trying different modes of transportation, finds that the most comfortable is the sleeping car. He can look at his journey and his goal in three different ways. First, he can become so attached to the sleeping car that he forgets the goal of his journey; and even if he arrives accidently at his goal, he does not understand that he has arrived at the place he had marked at the beginning as his destination. Second, he is not attached to the sleeping car, but since the train is going to a place that is not his goal, he renounces the goal of his journey and surrenders to tragic despair; or he begins to conduct little bits of business, goes to the market, and takes rides on the trolley and on suburban trains, though he keeps saying that it would be nice to travel to the place that he had originally marked at his destination. Third and last, he thinks of nothing but the goal of his journey, does not consider the inconveniences of travel, and keeps moving in one direction, without paying attention to the modes of transportation as long as they take him to this goal that lures him.

The first case brings to mind the rationalism encountered in Catholicism and in western European philosophy in general, its most extreme expression being the theory of Hermann Cohen, where the method is its own goal, so that the goal is ignored in order to ensure perfection of the method. If we think about the world in a certain way, we get a system that may be antireligious. That is the kind of thought we find, for example, in rationalistic Catholicism, although the living body of the Catholic Church contains various modes of thought, including profoundly vital ones.

The second case brings to mind the various types of positivism encountered in Protestant countries; this is a tragic work-a-day positivism, convinced that it is impossible to develop a system of spiritual knowledge. The work-a-day positivistic method of developing a world-view characterizes the Russian religious intelligentsia, the clergy, and part of the laity. Since this intelligentsia, in addition to being lazy, does not believe in the possibility of a rationalistic world-understanding, it has a good excuse to do nothing, that is, to avoid developing a world-view. This spawns different types of positivism, rationalism, sensualism—fragments of ideas assimilated unconsciously. The most dangerous aspect of this is that no thesis is apprehended consciously, and therefore no arguments are possible, since the thrust of any argument would hit an unconsciously assimilated proposition. This is the greatest evil of our life: there is no consciousness here of a great error having been committed, and repentance is therefore impossible.

The third case brings to mind the situation of a thinker when for him no method is a goal in itself, but every method can be either good or bad; when no method is accepted or rejected, there is faith in the goal. If faith is "the evidence of things not seen,"[1]

1. Hebrews 11:1.

we get a convergence of the objects of faith and the objects of knowledge, and the true method here is faith. And the means for discovering this method consists in the incarnation of faith in a concrete psychology. This is not eclecticism: the methods are not adjusted to one another, although, indeed, from the western European point of view there is a mixture here, a transition from one method to another, a change in the meaning of terms. We find an example of this is the writings of the Holy Fathers when they are considered from the point of view of western European thought; but this thought imposes demands on the Fathers which they would have considered unrelated to their task, and it does not notice the Fathers' direct, livingly concrete apprehension of the world by an act of faith.

When you try to explain something more rapidly, you will touch on everything, and the more diverse these objects, the more convincing your words will perhaps be, and every example will be more economical. The methods are unified and manifest the energy of the argument's creator. The Christian world-understanding can be compared with a poetic work: it is not a valid criticism to say that the thinker said something else previously, since, in essence, it is all the same. Consequently, the religious thinker's every term and method can elicit both yes and no. This kind of thought has a symbolic character, and if you have not understood this, you have not understood anything. Every attempt at rationalization is a distortion. The persuasive power of such thought is internal, not external.[2]

---

2. Marginal note: "The essence of religious thought is its symbolic nature. It is not a philosophical type of system. To accuse it of rational inconsistency is to misunderstand it completely. The unification comes from within." On a separate loose sheet in the notebook there is the text (abridged):

Christian thought has a teleological character. Not "why" but "for what" is the one goal, the point of reference, and not because we exert ourselves in all manner of ways to construct it in this way. We need to amplify the object of faith, and then it will appear with particular clarity in our consciousness. That is a sign of the weakness of our thought. Long motifs are needed only at the beginning, since short ones are so rich that we are not able to appreciate them, asking instead that they be unfolded for us. And then the mind (this word understood as the Holy Fathers understood it), as the center of the spiritual being, learns how to permeate into shorter and shorter formulas, which finally converge in the one most sweet Name of Jesus Christ. In Him, truth and being are one. "All things were made by Him."

Neither in the thought of the Church nor in the thought of the common people do we find a splitting into the subject and object of knowledge. This thought is real; it is symbolic. About everything one can say both yes and no. For western European thought the law of identity is obligatory. A is A, and A is not

"If all methods are taken in isolation, Orthodox thought might appear to be a type of eclecticism. That is the case from the point of view of rational thought. That is western philosophy's view of patristic thought. The argument begins using one method, and then it switches over to another method. A term is first used in one sense, and then it is used in another. One starts with certain premises but then changes them. This approach falsifies the patristic method of inquiry by attributing to the latter an investigation of problems it does not pose. The true method is the direct, living apprehension of an object by faith, and it can find sustenance in all manner of means. If we wish to explain something, we take comparisons from various domains, not just from one. This kind of thinking can be compared to poetic creation. To every proposition a religious thinker can say both yes and no. Religious thought is essentially symbolic. To attempt to transform it into rational philosophical thought is to misunderstand it completely. The unification here comes from within."

not-A. These laws of logic have no significance for concretely religious thought: ontologically, we have the affirmation that A is both A and not-A. How is it possible that a thing can be greater than itself? It contains both itself and something greater and higher than it. Here one can find the profoundest opposition between two types of thought: (1) religious, universal-human, ecclesial thought and (2) the deeply degenerate thought that calls itself "scientific" and that characterizes only a thin stratum of western European society, the cliquish thought around which western philosophy is oriented.

The use of personal motifs can often elucidate a discussion. I therefore wish to relate something from my biography.[3] When my world-view was being formed, the train of my thought was approximately as follows. I grew up believing in scientific thought. I did not know any other kind of thought, and I believed in it, that is, I recognized it as self-sufficient and self-evident and regarded as false everything outside of it. But then I began to notice in it a number of inconsistencies and cracks, this coinciding with a profounder experience of life, predominantly of a mystical character—and my mood turned black, like that depicted at the beginning of Goethe's *Faust*.[4]

---

3. On a loose sheet inserted in the notebook there is the following text (abridged): "The teleological character of thought. This goal is the object of our faith, which we wish to amplify and separate into individual elements. The necessity of such separation is a sign of weakness. Long prayer is needed at the outset, but then it is compressed into concise formulas. As the mind grows stronger, it can comprehend concise formulas. The final prayer is the Name of Jesus. The necessity of unfolding a world-understanding is a sign of weakness. For ecclesial thought both ontological and gnoseological Truth comes from one Word: 'All things were made by Him'."

4. Marginal note: "I experienced a crisis—consisting in loss of faith in the possibility of knowledge."

Around this time I encountered the works of Leo Tolstoy. But it was not their dogmatic side that attracted me. I held that if you're going to seek dogmas, you should, of course, seek them in the Church, not in Tolstoy; and of course I did not seek them in his works. I was especially drawn to his *Confession*. If one excludes the end, which is tendentious and was added later, this is a great work of apologetics, which should be distributed widely. It has the effect of an enormous explosion and immediately destroys any naively optimistic view of life, raising the dilemma: either one must find the Truth or one must die from thirst for the Truth, die not only bodily, but more deeply, metaphysically.

My disenchantment, combined with the influence of Tolstoy's works, threw me into a profound despair. This state continued for about a year. But my soul nurtured the secret belief that the Truth had to exist, that knowledge of It had to be possible; otherwise I would have to die. A spiritual instinct struggled within me, however, instilling in me the desire not to die (spiritually, of course). If knowledge of the Truth is necessary for life, and if hundreds of generations living before me had some contact with It, and since I could not be so conceited as to think that I alone would attain the Truth and that those millions of people had lived worse than cattle, it followed that either my search would be fruitless or those people did indeed possess something of the Truth. Meanwhile, since I could not accept that I would always be in this black hole, it followed that the Truth had always been given to people and that it is not the fruit of the study of some book, not a rational construct, but something much deeper living within us: It is the thing by which we live, breathe, and are nourished. The various means of expressing It can be valuable or they can be harmful; but they are only superstructures above

It; they are something secondary. Consequently, all the theories of earlier thinkers contained something of the Truth because humankind was made up of people, not of cattle, and I could not accept that I alone possessed the pure core of It, whereas they had only the appearance of It, the husk. And since all those theories were, on the one the hand, not a mere husk, while, on the other hand, they had only temporary significance, it followed that this, at the same time, was both husk and core, both garment and body—it, not it, and more than it. All these theories were symbols, true for their creators; while for others they were dead garments and therefore harmful.[5]

The notion of the symbol—every living world-understanding which is necessary for oneself, one's friends, and one's family, and not for the cabinet or rostrum, is symbolic.[6]

There can be no metaphysics external to the center of our life which could lead us to the Truth. Metaphysics must come from the Truth Itself; it must originate in our experience of the Truth, since one cannot obtain the Truth just by putting together factual material; and even if the Truth were to be obtained by some accident, we would not be able to recognize It. Methods are determined by the goal. And the possession of this goal differentiates us with respect to spiritual structure. The structure of our thought is determined by the goal for which we live; it depends on the structure of our spiritual life, on that center

5. Marginal note (abridged): "When this husk perishes, it in part expresses the Truth and in part it is a dead garment. It must be received inwardly, not outwardly. That which does not pass from heart to heart is a harmful, dead thing. The hole of death. . . . The greatest responsibility for one's life."

6. Marginal note: "Symbolism and not-symbolism."

toward which we are turned. The forms of the structure of our thought, the set of laws, constitute the phases of humanity and of our condition, which is transitory.[7] "Ye shall know the Truth, and the Truth shall make you free"; it will free you from enslavement to yourself: the objectivism of the laws of thought, the spirit knowing the Truth and looking at itself from the side, surpasses itself, and what before had seemed incomprehensible now becomes clear at another stage.[8]

There can be no self-sufficient metaphysics or science: they originate in the object of faith, which exists also in scientific thought.

One of the most important propositions of Christian metaphysics is the difference between energy and essence. Essence is that side of an object which is turned *ad intra*, while energy is the side which is turned *ad extra*. These sides are not opposed to one another, but are manifestations of one and the same object. Energy is the revelation of that which reveals. Etymologically, the Russian terms for "revelation" and "to reveal" (*iavlenie, iavl'at'*) have the sense of "not to conceal the essence," while the western European term "phenomenon" means the opposite.

We see the light of the sun, not the sun itself, but in this light we

7. Marginal note: "How is symbol possible in the metaphysical respect? From the ecclesial point of view there can be no abstract or external metaphysics; there can only be a metaphysics that comes from the Truth. It is impossible to obtain the Truth by taking some notions and putting them together in some way. One or another order of thought is only a phase, not the final stage,"

8. Marginal note (abridged): "We must speak of the laws of thought and being, which originate from our Faith in the Only Begotten Word of God. 'All things were made by Him'."

see the sun itself. "Light of the Light, in Your Light, we shall see the Light."[9] And the true essence of the sun, its being not exhausted by its light, is at the same time adequately revealed in this light. Not the Kantian mirage. In every beam of solar rays we see the whole sun, although the brightness may vary. Based on this adequacy of the manifestation of a thing to its energy, one can say that the energy of a thing is also the thing itself.

The essence of the Palamite disputes. Their general philosophical significance. The temptation of eclectic collecting.

The Christian world-understanding is not an eclecticism like that of Plotinus: the Church did not take anything from historical influences, but assimilated that[10] which enabled it to unfold its teaching to the maximal degree; for this reason she received the most material from kindred teachings, from universal-human, living movements. The Church adopted a great deal of her terminology from these movements.

9. From an Orthodox prayer.
10. Marginal note: "Teachings estranged from the universal-human are remote from Christianity. Platonism is an anticipation of Christianity, providing a great deal of the terminology used by the Church."

Lecture Nineteen

ONE OF THE most important questions is the question of the symbol. The entire Medieval world-understanding was symbolic. A symbol can mean any reality that contains in its energy the energy of another reality higher in value and hierarchy; the lower reality is a window into the higher reality, and if the lower reality is destroyed, the light of the higher one fades, not in itself but insofar as the window is closed. This means not that the higher reality has stopped existing, but that the window has been closed. The window is the sensuous thing through which the higher reality is revealed.

The lower receives its name from the higher, and the lower name is in second place according to the hierarchical value of the answer. The lower rebels against the higher. For example, myrrh is both fragrance and Divine Grace, which is inseparably united with the energy of this fragrance. Anointing with it is union with grace, and therefore about myrrh it can be said: "This is Divine Grace." But someone could object and say "it's just oil mixed with fragrance!" Formally, that is true; but it is not true metaphysically or ontologically, since in that case the higher value would be destroyed by the lower.

There was this incident: a peasant woman at the market was berating Jews, and I.V.P.[1] told her: "Why are you berating them? Don't you know that Jesus Christ was a Jew?" The woman answered: "What are you saying, sir? He is the Holy Spirit." This answer, too, is dogmatically correct; according to Gregory of Nyssa, for example, Christ, as well as the Father, can be called the Holy Spirit. Thus, both answers are correct; but when these two answers collide, one of them will be metaphysically false, for it is false to put a lower truth in first place.

Destructive attacks on the symbol can come, first, from rationalism; and, second, from naturalism: if one removes the whole sensuous husk from the symbol, then the spiritual content will disappear and the symbol will become invisible; on the other hand, it is possible to thicken sensuously the husk to such an extent that it will become impenetrable for the spirit and the spiritual will become invisible. Consider, for example, what happens when we try to justify scientifically a religious truth, e.g., the Ascension. When we try to answer the question how this was possible, the husks become iron-clad. This is not a legitimate question in the mystical order of things. The Kantian understanding is based on theses with a definite anti-spiritual value. And if this fact is denied, all of Christianity is destroyed; so we must accept the fact. The Kantian assertion is valid insofar as its premises are valid. If we accept and assert the Ascension, Kant and all his constructs will have to be cast aside. And they must be cast aside, since I have neither the possibility nor the desire to sort through these facts. I am saying that we will discuss this later, if we have any free time.

---

1. Ivan Vasil'evich Popov (1876–1938?), professor of theology and theological writer.

The fact of the Ascension is based on an infinitely significant and valuable fact—a fact of spiritual vision. Western art thickened the symbol and became religiously esthetic; and this husk stopped being a sign of a higher essence and the fortification of the spiritual side. This is particularly true in the case of Renaissance painting. As the epoch progressed, the husk became regarded as intrinsically true and beautiful.

The second distortion of the symbol is the rationalistic one. Here the attempt is made to remove totally the sensuous husk, to scrape out the symbol in such a way as to do away even with the paper on which its words are written. Thus there remains only the pure meaning, the forms of expression of which become more and more attenuated and, in the end, arrive at agnostic inexpressibility, since they wish to express the higher reality adequately, to give this reality and not just indicate it. This is scholasticism. *Ergon* and *energeia*. One wishes to transform life into *ergon*, to abolish energy, so that only a shadow or residue of reality remains, but one that continues to be just as sensuous in character.[2]

Scholastic theology continuously oscillates between these positivistic-realistic and agnostic-rationalistic distortions of the

2. Marginal note: "Such are the scientific propositions; they become pragmatic. In the domain of art, too, the symbol can thicken. The emergence of western European art is explained by the thickening of religious symbols, which gradually stopped being transparent and became valuable in themselves and autonomous. Through rationalism and agnosticism the attempt is made to remove everything but the pure meaning, the spiritual content, to scrape out the sensuous husk; but then we stop knowing the spiritual world, for one wishes to express it not symbolically, but adequately. This leads, among other things, to the emergence of a scholasticism that abolishes energy, leaving only a logical *ergon*."

symbol and becomes a non-spiritual world-understanding. For scholastic theology the icon is one of the incomprehensible symbols. For this theology the icon and liturgy do not exist; it regards them not as a process of world-understanding, not as concepts and categories, but only as an external thing that can be studied externally. In contrast, for the religious world-understanding the icon or spiritual countenance is a category, an organ of knowledge, not in the capacity of an example, but in essence.

I wish to discuss the concept of the icon at greater length. In accordance with the decrees of the Ecumenical Councils, the icon is an image reminding us of (or raising us to) the Proto-image. We venerate icons, burning incense before them, because the veneration passes on to the Proto-image. The terminology here is Platonic. The Proto-image is not a thought or "idea" conceived through the abstraction of a representation, but a spiritual reality, that spiritual essence which makes a saint himself, that power which forms his spiritual personality and organism, shines in his countenance, and is visible through the coverings of his body. The icon is, first of all, the energy of the saint's body and, secondly, that which the artist understands the saint's essence to be. When it gazes at the icon, the mind (in its patristic sense, meaning the center of the spiritual being) rises from the image to the Proto-image through remembrance, *anamnesis*, as the decree of the Ecumenical Councils says.

The word *anamnesis* is a term from the Mysteries and from Plato. It signifies our spiritual-mystical state which arises from the touching of the Platonic ideas; it is the mystical remembrance that is ignited when we touch the icon. I insist that this term has a metaphysical meaning here, not a psychological one. The icon reveals the Proto-image: when we touch the wood, the

[140]

canvas, and the varnish, we touch a certain spiritual reality. The icon is a window into another world, and as long as we do not see this window in the icon, our relation to it will be false, idolatrous. By faith we see through it as through a glass that is either dim or transparent, depending on the elevation of our spiritual state and on the degree of the artist's spiritual penetration. We have a spiritual vision; we see another world "as through a glass, darkly."

In antiquity, in the mystical era, the iconostasis did not exist, but when the spiritual consciousness began to fade, iconostases began to increase in size. In Russia in the 16th and 17th centuries the sanctuary is another world. This section of the building, the floor, the space, all of which can be sensuously built and destroyed, is a symbol that contains the energy of another world. Although the floor remains a floor and the paints remain paints, another world is given to us through them. In itself this place is full of awe. But people who do not feel the fear of God, the direct presence of God, must be separated from it by wooden barriers. The sanctuary, in the capacity of heaven, is guarded by angels and saints; and the iconostasis reminds us of their presence, reminds us not psychologically but ontologically and mystically: they are here. We have become so thick-skinned that we have to be reminded of this; we have to be struck on our mystical feelings before we can be awakened to the fact that the saints are here. The iconostasis is, so to speak, the materialization of spiritual powers guarding the sanctuary; it is a world subject to special laws.

But, starting with the Renaissance epoch, painting and all of life were rapidly distorted; and even today we are not set afire even by icons—for us they are a window that has grown dark.

I wish now to discuss the veneration of icons in relation to reverse perspective. That is the touchstone for any icon.

Ancient icons often display a crude distortion of the laws of direct and reverse perspective. Planar boundaries. Edges. Books are depicted contrary to the laws of linear perspective: the Gospel book displays edges on all four sides and from different points of view. Chambers can be seen from three sides. There seems to be a crude illiteracy here. Bodies are foreshortened and bounded by curves. The planes of the face and of the whole head are turned away; i.e., when a face is depicted in profile, the other half of the face is depicted too. Parallel lines diverge toward the horizon, etc. However, all these distortions of perspective do not provoke an unpleasant feeling, but are even agreeable, so that icons with a more correct perspective seem cold, soulless, and boring. On the other hand, we feel there is a rightness in icons with reverse perspective. But perhaps we like their naivete, their childishness? But these icons were painted by the greatest masters, and therefore are we ourselves not naïve when we call these icons naïve? And, truly, we find in them not happy accidents but a conscious underscoring of details: the floor is illuminated more brightly; on the Gospel book all four images are painted cinnabar, making this book the brightest place on the icon, and so on. They display not an imitation of nature but an originality of depiction. They also display a multi-centeredness, i.e., different points of view. We see a subtle artistic calculation in parts of the chambers and in the faces. Clearly, all these distortions are meaningful.

We have to clarify: Is perspective a true expression of reality? Or is it only a schema, one of the possibilities derived from a particular world-understanding; an orthography corresponding to the style of a particular age, the violation of the laws of which does, perhaps, as little harm as errors in the writing of a saint? In paintings of ancient Babylon and Egypt there is neither direct nor reverse perspective, but they are nevertheless characterized by a startling truthfulness bearing witness to a great power of observation. Why is it that these cultures did not notice perspective? Moritz Cantor[3] tells us that the Egyptians knew mathematics. They knew proportionality and scale. It is not known why they did not use linear perspective, but they did not think of a picture as a wall placed between the viewer and the action. Cantor says that this might have been for religious reasons. Art was closed up in itself. This might have been due to emancipation from perspective, to the fact that its period of maturity had passed, or to the fact that it had been rejected from the very beginning.

The Egyptian world-view and Egyptian art were not tied to a single point of view or to a particular moment of time. Individuality in applied art does not equal truthfulness of being. Anaxagoras and Democritus were the first to clarify the laws of perspective and scenography (Vitruvius), explaining how rays of light must be transmitted in order to pass from a picture onto the retina as from a phenomenon.[4] Thus, perspective was known in early antiquity. But it was not used because the task of painting was not to duplicate reality but to provide a more profound understanding of it. The stage decoration was not symbolic; it was a deception; whereas painting—the truth of life—

3. Historian of mathematics (1829–1920).
4. Marginal note: "Perspective arises in the domain of applied art—in stage design. Its invention is attributed to Anaxagoras in connection with the staging of Aeschylus' tragedies. The stage replaces life."

was symbolic and significant.[5] Stage decoration is a screen, whereas painting is a window into reality; stage decoration is an imitation of life: man is a prisoner chained to a cliff, to a single place, even to a single point. He is a single eye staring involuntarily at the stage. A living man is replaced by a spectator poisoned by curare which paralyzes the ability to move while leaving the consciousness unaffected.

Thus, perspective was known as early as the 5th century BC, and if it was not used, there must have been higher motives for this; given the knowledge of geometry then, it was not that perspective was not noticed, but that it was considered unartistic. The degree of the sophistication of mathematics at that time can be seen in the fact that Ptolemy compiled complex maps of earth and heaven that were approved and used as late as the 16th century. Painting with direct perspective developed from stage decorations. It extended the stage, destroying the wall separating us from the action. It developed in the age of the so-called *graeculorum*, the vapid elegant "Greeklings" who frequented Rome at the time of the Empire, spreading corruption and debauchery there. The paintings at Pompeii—this baroque of the ancient world—aimed to deceive the viewer. They were the works not of artists but of virtuoso artisans. Meanwhile, in the 4th century BC, perspective vanished.

I have obtained all this information from Alexandre Benois' *History of Art*.[6] But it can be found in any book on the history of art. It must be noted, however, that almost all of these authors

5. Marginal note: "Medieval art creates symbols, not likenesses."
6. *The History of the Art of All Eras and Nations*, Collected Works, vol. 4, Petersburg, 1912–1917.

engage in a reverse reasoning: they regard the period of direct perspective as a period of flowering, whereas it was actually a period of decline. That is understandable. Benois and the others reason thusly: how could it have been bad if it led to me? The highest praise he can offer is: "Our contemporaries cannot surpass this."

Lecture Twenty

18 November 1921

WHEN ILLUSIONISTIC ART evolves, when the need arises to create only an appearance of reality, and not a window into it—a means to deceive the immobile viewer emerges: perspective. Nineteenth-century man made himself the absolute measure of truth and beauty. He forced the whole universe to revolve around him. Perspective is the expression in painting of the Kantian world-understanding.

The Middle Ages are commonly reproached with the absence of an understanding of space. But what was absent then was a Euclidean understanding of space, an understanding that is reducible to perspective and proportionality and unconsciously asserts that it is self-evident that forms do not exist in the world, and that every form is placed like an empty cage on top of soulless material. By an irony of history the Renaissance world-view, which started by proclaiming the rights of man and of nature, came to reject them. The goal of this world-view is not grateful acceptance of all reality, not creative activity, but the creation of simulacra and life among them. Man knows that in order to desire it is necessary to be real and to base oneself on reality; and space is not a graph but reality itself. Contemplatively creative culture and rapaciously mechanical culture.

The divergence of parallel lines toward the horizon (and so on) is due not to ignorance but to the artistic method. Children's drawings with reverse perspective resemble Medieval ones, since children's thought is not naive but of a special type. The Renaissance began with the "little lamb" of Francis of Assisi.

As soon as he had the urge, Giotto almost immediately discovered perspective. Fresco is not a decoration, but an independent kind of painting. But even Giotto, though he knew how to use perspective, rarely had recourse to this technique. Only in the 16th and 17th centuries, when suspicion was cast on the absoluteness of theocentrism, did attempts evolve to replace Divine reality with simulacra, with theater. Giotto's landscape derived from the decorations of the Mysteries.

Perspective does not accord with man's nature—proof of this is the complex and prolonged labor that was required to hammer out the notion of perspective. The period between the 15th and 18th centuries witnessed a forced retraining of human psycho-physiology. Perspective is alien to man because it is a conspiracy against natural perception: it demands that one see not that which exists but that which is desirable to see.

Distortions of perspective are connected with the high qualities of icons. Historians of culture put a greater value on a cultural phenomenon if it approximates phenomena of the second half of the 19th century. Perspective is an expression in painting of the Kantian world-understanding. A picture of the world with perspective is not a fact of perception but a requirement in the name of certain principles.

General Conclusion
After the Course

Which Was Not Completed

IN PERSIA there is the sect of the Yazidis that worships the principle of evil in order to propitiate it, while neglecting the principle of good, since it has nothing to fear from it. In our Russian world-view too, the Manichean point of view must be eradicated. We must reeducate ourselves, destroy the hard shell around us, and come to understand that the world is a symbol and full of symbols and that therefore our world-understanding must be symbolic. Yes, it "must" be, but I do not see any real evidence that such a thing is happening.

An ecclesial brotherhood. The Lavra of St. Sergius. Plato's Academy. The Pythagorean School. The Priestly Collegium of Egypt. What they have in common is that they are not a school but an atmosphere of comradeship with vital personal ties connecting more tightly than ties of family. Only organic closeness can create a world-understanding, since it is life itself that educates us here in the fulfillment of that which is objectively necessary. This must take place in separation from the world. An element of the monastery is necessary. This does not mean that one should leave one's employment; it means that groups of families must be separated from others to keep those others from dissipating their energies. Our monasteries suffer from a lack of ecclesiality.

Appendix One

Transcript of a Conversation that Father Vasily
Nadezhdin, A.N. Sokolov, and the Unknown
Author of the Transcript had with
Father Pavel Florensky
in November 1921

SOMETIME AROUND November 1921, Father Vasily F. Nadezhdin, Alexander N. Sokolov, and I were sitting with Father Pavel Florensky in the vestibule of the church of Petrov Monastery, awaiting the end of the preceding lecture. In the course of our conversation Father Pavel asked if any of us had ever noticed which of his hands was warmer. None of us could answer this question. He said he was asking us this question in order to confirm his observations. "Because I have noticed," he continued, "that under ordinary circumstances, that is, when I am sitting calmly or lying in a room where the temperature is stable, my right hand is warmer than my left. But when I am praying, and especially when I am performing the liturgy, my left hand is warmer than my right. Have you ever noticed this?" he asked Father Vasily. "No," answered Father Vasily. "My right hand is always colder, because my church isn't heated and I tend to wear a glove on my left hand." The conversation then took another turn.

That evening Father Vasily told us that he had received a letter from Father Ivan Kozlov of the Chembarsk District of Penzensk Province, asking that Father Pavel give his opinion concerning incidents of the kind described in the sermons of Grigory Diachenko: namely in Moscow sometime in the 17th century, people began to hear knocking, a loud racket, and other disturbances coming from an almshouse. The one knocking called himself by the name of a Boyar's son who had recently died. It was noticed that all these spiritual shenanigans were absolutely in keeping with the character of this Boyar's son, for he had been very unruly. How should one view these incidents? This is what Father Pavel said: "Final death comes to a man much later than the moment of death attested by the physicians, for a man has a number of other bodies besides the body in the strict sense of the word, the physical body; he also has an etheric body, an astral body, a mental body, etc. Final death comes around the fortieth day, when (according to the Holy Fathers) the heart decays. So that when a man dies, his astral shells remain in the world and can be possessed by evil spirits who, so to speak, galvanize the astral corpse, much in the same way that it is possible to send one's energy into a corpse and make it move (as in the case of the false vivification of corpses by yogis). The evil spirit thus has reason to call itself by the dead man's name. We therefore see in these shenanigans the manifestation of a man's character and that his corpse is an agent acting under the influence of outside powers."

Appendix Two

From Father Vasily Nadezhdin's Notebook

Entry on 4/17 November 1921[1]

REMEMBER to ask Father Pavel:

Is Christianity compatible with capitalism? How great is the role of the Masons in the Russian Revolution? How should one understand Christ's "thou shalt not kill" in the context of our present reality? And His "judge not"? Is capital punishment compatible with Christianity? Are Christian civil courts possible, or is jurisprudence a pagan element? Is it correct to say that neither servants nor slaves should exist in Christianity?

---

1. Nadezhdin was one of the auditors of the lectures and his notes have survived.

Index

Made in the USA
Middletown, DE
19 January 2018